Space Invaders

Radical Geography

Series Editors:
Kate Derickson, Danny Dorling
and Jenny Pickerill

Also available:

In Their Place
The Imagined Geographies of Poverty
Stephen Crossley

Space Invaders

Radical Geographies of Protest

Paul Routledge

www.plutobooks.com

First published 2017 by Pluto Press
345 Archway Road, London N6 5AA

www.plutobooks.com

Copyright © Paul Routledge 2017

The right of Paul Routledge to be identified as the author of this work
has been asserted by him in accordance with the Copyright, Designs
and Patents Act 1988.

British Library Cataloguing in Publication Data
A catalogue record for this book is available from the British Library

ISBN   978 0 7453 3629 9   Hardback
ISBN   978 0 7453 3624 4   Paperback
ISBN   978 1 7868 0110 4   PDF eBook
ISBN   978 1 7868 0112 8   Kindle eBook
ISBN   978 1 7868 0111 1   EPUB eBook

This book is printed on paper suitable for recycling and made from fully
managed and sustained forest sources. Logging, pulping and manufacturing
processes are expected to conform to the environmental standards of the
country of origin.

Typeset by Stanford DTP Services, Northampton, England

Simultaneously printed in the United Kingdom and United States of America

Contents

Figures

Acknowledgements

This book is the product of a host of relationships, collaborations and inspirations over the past 30 years. It would not have been possible without the love and support of Teresa Flavin, the keen eye of Dave Featherstone, the pleasures of collaboration with Andy Cumbers and Kate Derickson, the strange beauty of Pollok Free State, the word power of Narmada Bachao Andolan, the internationalism of People's Global Action, the direct action of the Bangladesh Krishok Federation, the joyous rebellion of the Clown Army and the creativity of Climate Games. I also want to thank my editors and my once and present colleagues in geography at the Universities of Glasgow and Leeds for the many discussions, comments and reflections that have improved my work over the years.

Series Preface

The Radical Geography series consists of accessible books which use geographical perspectives to understand issues of social and political concern. These short books include critiques of existing government policies and alternatives to staid ways of thinking about our societies. They feature stories of radical social and political activism, guides to achieving change, and arguments about why we need to think differently on many contemporary issues if we are to live better together on this planet.

A geographical perspective involves seeing the connections within and between places, as well as considering the role of space and scale to develop a new and better understanding of current problems. Written largely by academic geographers, books in the series deliberately target issues of political, environmental and social concern. The series showcases clear explications of geographical approaches to social problems, and it has a particular interest in action currently being undertaken to achieve positive change that is radical, achievable, real and relevant.

The target audience ranges from undergraduates to experienced scholars, as well as from activists to conventional policy-makers, but these books are also for people interested in the world who do not already have a radical outlook and who want to be engaged and informed by a short, well written and thought-provoking book.

Kate Derickson, Danny Dorling and Jenny Pickerill
Series Editors

1

Radical Geographies of Protest

*Spatial Strategies, Sites of Intervention
and Scholar Activism*

Protestors are space invaders. In the course of protests, all kinds of spaces – such as homes, corporate offices, streets and factories – are used, occupied, defended and abandoned. Particular places provide protestors with opportunities and constraints as they wage their struggles. Places can influence the character of protests as well as being transformed by them. Protestors make space, and in so doing they can imbue places with different meanings and feelings. In short, protest always has a geographical character and this has implications for the emergence, character, impact and outcomes of particular struggles.

Protests form part of a broader set of interactions, repertoires and processes that are termed 'contentious politics', which can include strike waves, revolutions, armed conflict, civil wars, guerrilla insurgencies and democratic processes involving political actors and governments.[1] Protests are prosecuted by a spectrum of different societal actors including individuals, groups and, of particular interest to this book, social movements – that is, organisations of varying size that share a collective identity and solidarity, are engaged in forms of conflict in opposition to an adversary (such as a government or corporation), and attempt to challenge or transform particular elements within a social system (such as governments, laws, policies, cultural codes and so on).[2]

Whether it be the alter-globalisation mobilisations of the turn of the century, the flurry of Occupy protests that peppered the planet a few years ago, the recent wave of anti-austerity mobilisations or ongoing protests against the construction of dams or the spread of agribusiness, there is a geographical logic to all forms of protest. Through a discussion of different case studies, I will explain how an understanding of 'radical' geography – an approach to geography that is motivated by concerns for social and environmental justice within a global capitalist economy

– enables us to make sense of protests around the world, and provides a series of geographical strategies of use to protestors.

This book will consider two distinct yet interrelated geographical logics that are critical in overall strategic approaches to the prosecution of protest: a primary logic of *spatial strategies*, by which the character of protest is informed by, and shapes, the geographical contexts in which it takes place; and a secondary logic concerning key *sites of intervention*, physical and conceptual targets within a system that are directly related to a protest's concerns, goals or broader strategies. Taken together, these 'logics' provide an innovative approach to protest that enables us to understand why such mobilisations occur where they do, and provides useful insights for students and activists wishing to make sense of the world of protest and build effective campaigns.

In this chapter I will discuss what is meant by geography, and in particular, radical geography and the contributions that it has made to social movement theory. I will also consider the practice of scholar activism, and issues of ethics and representation concerning collaborating with and writing about political struggles. Following a discussion of politics, protest and power, I will introduce six spatial strategies and nine sites of intervention that I use to interpret political protest from a radical geographical perspective.

GEOGRAPHY AND THE RADICAL IMAGINATION

Human geography is concerned with the people and places that make up the world, their similarities and differences, their connections (or lack thereof) and the processes by which our world is structured into identifiable places and peoples. Two key geographical concepts used throughout this book are place and space. Place refers to a particular geographical locale distinguished by the cultural or subjective meanings through which it is constructed and differentiated (from other locales). These meanings can change over time, and places are always connected to other locales regionally, nationally and internationally through flows of people, investment, ideas, products and so on. Space refers to the ongoing flows, forms and social relations of the world in which we find ourselves. Space is never static, but rather plural, multiple and subject to transformation.[3]

At root is the recognition that everything happens somewhere, and that, for geographers, this is important. For example, where we are in the world is fundamental to what we see (or do not see), what problems we face, what languages we speak and think in, what we do, what chances we have in life, who we interact with and so on. In the language of human geography, we say that human and non-human things exist in and through space: in very mundane ways, all of us live in homes that organise the world into private and public realms; we live in settlements that we name and categorise by whether they are villages, neighbourhoods, towns and cities and so on; we belong to and reside in specific territories (such as nation states) that are differentiated from others in particular ways; and we move between places and across territories due to work, leisure, migration and so forth. In other words, spaces have a material reality and a symbolic significance that are important to how we experience and engage with the world.[4]

In addition, the various processes that generate social differences and inequalities (regarding health, gender, race, class and so on) are a product of how power and resources are distributed, manipulated and struggled over, which themselves are geographical in character. Such disparities have real impacts on geographical processes such as the location and provision of schools and hospitals (and hence access to doctors and education), transport provision (and hence people's journey times) and the location of waste. In short, geographical (or spatial) patterns produce, and are produced by, social relations and socio-economic processes.[5]

Radical geographers are not interested in merely mapping such differences within or between regions, cities and so on, but rather in investigating why and how 'context' enables an explanation for such differences. Such a geographical interpretation of social life enables us to begin to understand the reasons for the emergence of protest in particular places. For example, the environmental justice movement began in Warren County, North Carolina, in 1973, after the decision by the US state government to build a landfill for contaminated soil following the dumping of 31,000 gallons of polychlorinated biphenyl (PCB) by the Ward Transformers Company on the side of roads in fourteen counties in the state. The location of the landfill was Shocco, a predominantly poor rural town in Warren County whose residents were predominantly African American, with neither a mayor nor a city council. Residents feared that their groundwater would be contaminated by the toxic waste and so local community leaders organised protests against the construction of the landfill. Their protests attracted the support of civil rights groups across the nation and focused national attention on the interrelated issues of class, race and lack of political representation, and how they influenced environmental policy-making.[6]

Radical geographers are also committed to challenging relations of power and oppression in order to construct more socially and environmentally just ways of being and living. This means that radical geography has a particular interest in understanding practices and processes of social conflict and change, often in the form of social movements such as trade unions, farmers' movements and environmental justice movements. At times, it also means that radical geographers are involved in these processes and practices themselves. At root, radical geography is concerned with the *where* of protest and how this influences or shapes the dynamics of contentious politics.

CONTENTIOUS POLITICS, SPACE AND POWER: SOCIAL MOVEMENT THEORY AND GEOGRAPHY

Since the 1960s, the majority of social movement research that has emerged out of Europe and North America has focused on four key themes.[7] First, there has been research into the mechanisms by which the resources necessary for collective action are mobilised. Resource mobilisation theorists[8] have investigated the availability of organisational and personal resources, the importance of leaders as political catalysts and the role of interpersonal and inter-organisational networks in the circulation of resources and the creation of solidarities.[9]

Second, there has been interest in the relationship between structural changes and transformations in patterns of social conflict. Research on 'new social movements' explored how feminist and environmental movements reflected shifting concerns in society away from the production of material goods (the traditional focus of trade unions and other class-based forms of collective action) to the production of knowledge.[10] This also implied the construction of new forms of social conflict that focused on issues of cultural and political identity as well as the use of new technologies in protests.[11]

Third, there has been research into the role of cultural representations in enabling collective action. This has investigated how processes of symbolic production and identity construction have acted as key frames by which collective action is interpreted and motivated.[12] Since the 1990s, the role of emotional responses in collective action has also been examined.[13]

Fourth, 'political process' research has investigated the effects of the political and institutional context on social movements' development and evolution, focusing on the degrees of openness of formal political access, the degrees of (in)stability of political alignments, political conflicts within elite groups and the availability of potential allies.[14]

While these different approaches to understanding social movements have often implied the importance of geography, they have rarely explicitly investigated the implications of a spatial perspective.[15] Therefore, over the past 25 years, geographers have begun to address this lacuna by considering issues of place, spatial inequalities, networks and scale, and everyday spaces of activism.[16]

The Politics of Place

It is within particular places where everyday politics is practised and made real. This is because the processes undertaken by macro-scale institutions such as governments are translated into being in particular places. Because different social groups endow space with an amalgam of different meanings and values, particular places frequently become sites of conflict, where, for example, government policies are contested and reworked by social movements.[17] This is most evident in instances where different ethnic or nationalist groups contest the same political space (for example, Israel/Palestine).

Places can have a key role in shaping the claims, character and capacities of social movements, giving rise to particular 'terrains of resistance'.[18] Social movements frequently draw upon local knowledge cultural practices and vernacular languages to articulate their grievances. Indeed, particular activisms are embodied in and through particular places. This plays a distinct role in shaping both the political claims of actors and the perception of political opportunities, or what have been termed 'place frames'.[19] For example, the Zapatista insurgency (see Chapter 5) was informed and framed by the place-specific political and cultural economy of indigenous Mayan people in the Mexican state of Chiapas. Zapatista activists wear black ski masks, not only because the mask is an important cultural symbol in Mexico, but also because ski masks act to hide activists' identities from the state and symbolise the fact that Mayan indigenous people have been historically invisible to the ruling elites of Mexico.[20]

Spatial proximity between activists also enables strong social and cultural ties to be established (through trust and kinship networks, for example, and through shared language and traditions), which can then be drawn on to enable collective action.[21] For example, trade unions in Canada and the UK have recognised the importance of developing ties with other social movements in the places where they operate, generating 'community unionism' that addresses low levels of unionism and workplace protection for home workers.[22]

Having said this, it is important to recognise that places are the product of greater and lesser flows of people, capital, technology, cultural artefacts, commodities and media. As Doreen Massey has argued, 'each place is the focus of a distinct *mixture* of wider and more local social relations'.[23] As a result, the politics of place should not be thought of as somehow neatly bounded off from broader processes within society. For example, spatial imaginaries – that is, individual and collective cognitive frameworks constituted through the lived experiences, perceptions and conceptions of the world around them – mediate how activists evaluate the potential risks and opportunities of joining social movements.[24] They are also at work in transgressive political practices that challenge everyday understandings of places and frame certain protestors and their activities as 'out of place'. For example, in Argentina during the 1970s and 1980s, the *Plaza de Mayo Madres* (also known as 'the mothers of the disappeared') conducted mass popular protests against the 'disappearance' of family members during the military junta that ruled the country from 1976 to 1982. Motivated by anger, sadness and a clear sense of injustice, the protests took place in public spaces such as public squares and streets. The traditional understanding of such spaces in Argentina was that they were non-political (since proper politics was meant to take place in government buildings) and masculine (since public space was more associated with male activity – such as working or commuting to work – while women's activity occurred in more private spaces such as the home). These traditional understandings of space were challenged when the movement mobilised in public squares across the country.[25]

Spatial Inequalities

Economic and political processes occur in geographically uneven ways that produce variations in the grievances, available resources and development of social movements.[26] For example, the uneven urbanisation process within the Southern states of the USA resulted in the concentration of organisational resources (such as churches, people, money, social networks) available to African Americans involved in the Civil Rights Movement in only a handful of urban centres.[27]

Geographical variations in the relationship between states and civil society actors are important in understanding the context from which social movements emerge. Social movements are confronted by more or less democratic political systems, and this can influence the political opportunities available to them.[28] For example, trade unions are still accepted as legitimate 'social partners' in much of Western Europe, though they have been under attack in the United States, the UK and Australia, and are heavily censored or state-dominated in parts of Asia and Eastern Europe. Within countries, variations are also evident. For example, in the United States, the American Federation of Labour (AFL-CIO) has faced a more favourable organising environment in Northern states than unions in Southern states because in the former, unions were accepted as legitimate social actors and had long-standing traditions of local union membership and organising.[29]

Networks and the Politics of Scale

Social movements are networks of people, resources and connections. Most operate at the intersection of a series of overlapping scales – from more local municipalities, through regions to the nation state and, increasingly, international forums. These different politics of scale – and their associated networks of activity – provide movements with a range of opportunities and constraints.

The role of the internet in activist networks has been discussed widely.[30] For example, studies have been made of how national and international connections require activists from different struggles to negotiate between differently placed cultural identities, interests and imaginaries.[31] The identification with particular places can be of strategic importance for the mobilisation strategies of movements. Activists may deploy symbolic images of places to match the interests and collective identities of other groups in other places, and thereby mobilise others in terms of a common cause. Hence the ties to particular places can be mobile, appealing to, and mobilising, different groups in different localities.[32]

Nevertheless, movements that are local or national in character derive their principal strength from acting at these scales rather than at the global level. For example, transnational corporations such as Nestlé, McDonalds and Nike have usually been disrupted primarily due to the efficacy of local campaigns.[33] Where international campaigns are organised, local and national scales of action can be as important as international ones. For example, between 1995 and 1998, the Liverpool dockers went on strike to demand the reinstatement of colleagues who had been sacked. Their international campaign was instigated at the grassroots level and coordinated and operationalised by dockers beyond the UK working within established union frameworks.[34] Movements also utilise political opportunities at one scale to create opportunities at others. For example, trade unions have employed a range of multi-scalar spatial strategies to transcend national-level organisational constraints in order to challenge multinational corporations through lobbying key global institutions such as the World Bank and IMF.[35]

However, international alliances have to negotiate between action that is deeply embedded in place – that is, local experiences, social relations and power conditions – and action that facilitates more transnational coalitions. These may generate uneven political connections between activists and different types of geographical outcomes. For example, People's Global Action, an international network of social movements (discussed in Chapter 6), had to negotiate different understandings of gender and ethnicity between European and South Asian movements while attempting to forge an effective alliance against the processes of neoliberal globalisation.[36]

Everyday Spaces of Activism

Everyday practices of social reproduction such as cooking, knowledge-sharing and childcare have helped to mobilise, enable, resource and give shape to protests.[37] Particular infrastructures and practices are required to reproduce everyday life, such as food supply, shelter, sanitation and the maintenance of communal and private spaces, in protest camps such as those against road-building discussed in Chapter 3.[38] In addition, concerns over social reproduction can generate new spaces of activism in which women engage in activities in professional, public and private spaces.[39] Such concerns often necessitate under-standing the emotional repertoires and motivations of movements and activists and how these are implicated in spaces of activism in different ways.[40] What has informed much of this geographical theorising has been a commitment by radical geographers to the practice of scholar activism.

EMPOWERING RADICAL GEOGRAPHY
THROUGH SCHOLAR ACTIVISM

Geographers' concern with issues of social justice dates back to the 1970s and has been particularly concerned with active participation in different forms of activism.[41] Feminist geographers in particular have considered the ethics and power relations generated through such participation, recognising that scholar activists have social responsibilities – given their training, access to information and freedom of expression – to make a difference 'on the ground'.[42]

Scholar activism has been concerned with a range of issues. First, with developing practices aimed at social transformation, such as jointly producing knowledge with social movements to produce critical interpretations and readings of the world that are accessible, understandable to all those involved and actionable.[43] For example, the People's Geography Project organised at Syracuse University, New York, has attempted to make research and geographical concepts relevant to social struggles 'for' (and to an extent 'by') the people, such as through the Syracuse hunger project.[44] Second, with developing a politics of affinity with social movements, the connections and solidarities forged being a key part of activist research.[45] Third, with participating in the life-worlds of research subjects, and/or the participation of those research subjects in the production of geographical research.[46] Fourth, with actively contributing to the building of collectively organised non-capitalist spaces.[47] Finally, with considering how activist research can exist in relation to the neoliberal university.[48]

My scholar activism has been motivated by the belief that critical thinking and critical practice are mutually constitutive: they inform and produce one another. Sites of knowledge production occur as much at the grassroots – created by social movements, indigenous peoples, farmers and so on – as they do within academia. Indeed, many of the theories and concepts that have gained currency within academia owe their origin to grassroots community activists.[49] What is important is that such knowledge production (or co-production between activists and academics) is an accountable and reciprocal process.

In my work I have been committed to participation in, and collaboration with, a range of different movements, campaigns and protest initiatives, many of which are discussed in this book. In so doing, I have attempted to practise 'situated solidarity'.[50] This involves being emotionally moved to collaborate with activists and movements because of core values – such as those concerning dignity, self-determination, justice – that I share with them. Situated solidarity also implies the recognition of various ethical concerns – for example, that the production of knowledge and hence my representations of events (including those in this book) are partial. They have been influenced by my reception and interpretation of information, the quality of the affective link generated between my collaborators and myself, by the workings of my memory, and by my emotions, subjective evaluations and personal limitations.

Further, the ethical concerns central to situated solidarity necessitate an acknowledgement that economic, political and institutional processes and structures shape the contexts of research (and the practices of solidarity by scholar activists) as well as its effects. As a result, I have had to confront and negotiate the unequal power relations that exist between my own society and those in which I have conducted my work, as well as those between my collaborators and myself. For example, I enjoy a range of privileges that accrue to a (white, male) academic employed at a UK

Figure 1.1   Scholar activism involving activists from the Assembly of the Poor, Chiang Mai, Thailand, 2004. Photograph taken with the author's camera.

university – for example, financial (funding) resources and the ability and time to travel – that are frequently not enjoyed by those with whom I have collaborated.

While recognising that these ethical and power relations cannot be ever fully resolved, I have attempted to practise forms of situated solidarity that challenge traditional roles, hierarchies and the general order of things, and that involve attempts to disperse power away from academia towards community activists in the form of friendship, cooperation and empathy. Where appropriate, I have also tried to challenge activist norms and assumptions, such as those concerning gender relations within social movements. It is within the spirit of situated solidarity that this book has been written. An understanding of the interrelationships between politics, protest and power brings me to the geographies of contentious politics.

POLITICS, PROTEST AND POWER

The history of capitalism can be summarised as a process of conversion of common resources into private property invariably through violent acts of appropriation. Dispossession – for example, through the exploitation and plunder of resources from indigenous people or the forced removal of peasants from the land and that lands' enclosure by elite capitalist or state interests – is a constitutive feature of the social relations of capital as well as colonialism. Karl Marx termed this process 'primitive accumulation', a process by which the means of production is privatised so that those conducting dispossession can make money from the surplus labour of those who, lacking other means, must work for them.[51]

Over the past 30 years, this process has intensified, and globally we have witnessed increasing primacy given to market forces within national and international economies, the selling off of public services and assets to private enterprises, and an increasing inequality gap between the extremely wealthy and everyone else. The term given to the capitalist economic theory that has dominated the process that we call globalisation is 'neoliberalism', and it has been actively promoted by international institutions such as the IMF, World Bank and World Trade Organisation (WTO), powerful national governments (such as the United States) and transnational corporations.

The radical geographer David Harvey has argued that what he terms 'accumulation by dispossession' lies at the heart of this capitalist project.[52] By this he means that the processes of 'primitive accumulation' are still with us: the accumulation of profits and wealth – by businesses, banks, transnational corporations – is predicated upon the dispossession of people, communities and even whole countries of their resources and assets.[53] This has taken the form of increased privatisation (of public utilities, public institutions, seeds and genetic material, for example), financialisation, through the deregulation of the global financial system (such as speculation, corporate fraud, the raiding of pension funds), the management and manipulation of financial crises (such as contemporary austerity policies following the 2007/8 financial crisis, and the bailing out of private banks with public money) and state redistributions (for example, the privatisation of social housing, health and education, and the reduction of state expenditure on social welfare).

Such processes of appropriation and dispossession have always been met with resistance – confrontations between those who seek to privatise and make profit from these resources and territories, and those who would seek to use them for the collective or common good. Many of the case studies in this book are examples of such resistance.

Dominant neoliberal economic doctrine over the past 30 years has caused profound damage to democratic practices, cultures, institutions and imaginaries. Political participation and the right to equality have been reduced to market freedom and the right to compete, while individual activity in the market has replaced shared political deliberation and rule.[54] This has resulted in a 'post-political consensus' that has circumscribed what is deemed as 'legitimate' behaviour in institutions, the media and decision-making in society, as well as attempting to circumscribe what can be seen and heard in political debate and protest.[55]

Reflecting on this, Jacques Rancière argues that the social world and the people in it are constructed in particular ways that grant participation (such as in decision-making) to some, while excluding and separating others. This occurs through a set of formal and informal rules and procedures that determine what is visible and what can be said and heard in political discourse (as the 'normal' state of things), as well as in the spaces in which this occurs, and obscures and renders unrecognisable substantial portions of the population.[56]

However, Rancière argues that such rules and procedures are always incomplete and unstable, and thus open to challenge, when those excluded by such orders make themselves seen and heard by enacting political interventions. For Rancière, politics is characterised by 'dissensus': the appearance of subjects (women, workers, farmers, indigenous people) in a refigured space so as to be seen and heard in it. The practice of 'dissensus' places one world within another, if only momentarily, and makes visible the partiality of the political order. Politics in this sense 'confronts the blindness of those who "do not see" with that which has no place to be seen'.[57]

However, what is missing from the analysis of political practice by Rancière is a fully rendered account of the spatial politics of protest. To do this I will first discuss the work of the geographer Eric Swyngedouw, who discusses the staging of politics. Second, I will discuss the varied power dynamics involved in protest. Finally, I will draw upon this and other research in geography to develop a strategic framework with which to interpret practices of resistance.

Drawing on Rancière's work, Swyngedouw argues that to express protest against the political order requires the performative staging, in particular places, of practices of equality, freedom and so forth. In other words, alternatives to the current order have to be enacted. These practices simultaneously call into question the structuring principles of the established political order, making visible the inequalities and lack of freedoms (or wrongs) inherent in it. Over time, the intention is to extend the reach of these placed 'moments' of antagonism to more generalised political demands. In so doing, political activity is that which transforms bodies and places from that which they have been assigned by the political order. Through practices of resistance, spaces become political because they embody (and make visible) challenges to what can be said, thought and practised.[58]

Swyngedouw provides the example of the actions of Rosa Parks, who in 1955 in Montgomery, Alabama, sat in the 'wrong' seat on a racially segregated public bus, simultaneously staging equality (her right to sit anywhere on the bus) and exposing the inegalitarian practices of racism in the United States. This geographically specific practice was subsequently universalised across the United States through the Civil Rights Movement, where the placed moment of antagonism by Rosa Parks became the stand-in for a generalised democratic demand: the end to racial segregation and racial inequality. If Rosa Parks had taken a 'proper' seat on the bus, she would have remained invisible. By refusing to do so, she performed a visible act of equality that cut through the normal state of affairs of the US polity at that time.[59]

To be seen and heard in such contexts requires protestors to enact power and make power visible. According to the Italian communist Antonio Gramsci, the dominating power of the state and elite-class interests over citizens in society – what he termed 'hegemony' – requires both coercion (through laws, the police and so on) and the consent of citizens to that power. As is argued by the strategic non-violent theory of power (but is equally applicable to other resistance strategies such as armed struggle), all forms of dominating power depend on the consent and cooperation of those over whom power is wielded; in other words, a government's power depends on people's consent.[60] When people cease to consent, they enact protest, challenging dominant ways of doing and thinking in society, registering their values, beliefs and commitment, and announcing that alternatives are possible. In short, they render dominating power visible, open to confrontation, negotiation and possible change.

When protestors assemble on streets and in squares, they are exercising a plural and performative right to appear.[61] Zechner and Hansen argue that social power is made manifest when protestors become capable of creating certain effects in a given context. This involves the operation of four important forms of power. First, relational power, which is the terrain of individuals and groups engaging in temporary, ad hoc relations and encounters, where the spaces of social media and loose networks work to create calls for public mobilisations.

Second, compositional power, which is the terrain of the creation of common spaces of activity and protest. These include climate camps, popular assemblies and land occupations. In such spaces social relations can deepen and resources can be produced, stored and distributed. The imaginaries and slogans generated in these spaces of embodied experience shape and strengthen the ideas that circulate in social media. Third, organisational power, where the capacity to mobilise and organise is channelled into the terrain of formal organisations (such as social movements) or networks that produce, distribute and manage material and immaterial resources and symbols. Such power generates specific protocols and demands that engage with the state in attempts to effect change.

Fourth, representational power, which is the terrain of political parties, institutions and the media. This power is frequently in opposition to other powers, but nevertheless has an important relationship to them. This is because if relational, compositional and organisational powers are at work together and engage – through opposition and/or negotiation – with representational power, then the possibility for institutional transformation is increased. Social power is articulated across all four terrains: relational, compositional, organisational and representational. They constitute four interacting spaces of engagement that will be drawn out in the case studies discussed below, although their relative contributions in particular struggles varies considerably.[62]

Taken together, these forms of power can constitute 'popular power', that is, the capacity of marginalised folk to organise and coordinate structures to govern their own lives, thereby creating new social relations and forms of collective organisation. For example, since 2013 in Venezuela, 44,000 communal councils have been established in parallel to elected representative institutions whereby citizens engage in direct democratic participation in decisions that affect their lives.[63]

However, as I shall discuss in this book, popular power is rarely a pure space of resistance that sets itself against some form of totalising dominating power. Rather, practices of resistance are entangled in various ways with those of domination. For example, social movements frequently suppress their own internal differences in the interests of some broader strategy; hence resistance against the Vietnam War in the United States during the 1960s was criticized for its sexism. Meanwhile, the state – as a particular manifestation of dominating power – may well uphold certain democratic rights within a society that are being challenged by (anti-democratic) resistances such as the upholding of reproductive rights by various state legislatures in the United States in the face of anti-abortion opposition groups that have entailed the fire-bombing of reproductive health clinics.[64]

Through the staging of social power that renders protestors visible and listenable, and that calls into question the political order, protests fashion their own geographies of scale and expression. Below, I set out a framework of how this is achieved. This framework sets out a distinctive and innovative strategic approach to understanding the geography of protest that draws upon but extends work conducted by geographers on the spatiality of contentious action.[65] Through examining six different yet interrelated *spatial strategies*, I have attempted to consider the importance of place, scale, mobility, creativity, emotion and discourse to the practice of protest while also examining the power relations involved in its prosecution. I interweave this analysis with a consideration of the deployment of nine different yet interrelated *sites of intervention*:

material or conceptual spaces within a system where activists apply pressure in order to disrupt its functioning or argue for change as part of their broader strategic goals of campaigns.

My strategic approach to understanding the geography of protest is grounded in my research and political practice as a scholar activist in different geographical contexts over the past 30 years. As such, each chapter includes case-study materials drawn from my scholar activism (referring the reader to my earlier research in the endnotes), and also extends the analysis to include a variety of contemporary acts of resistance that exemplify the key spatial strategies and sites of intervention that I discuss in this book.

Spatial Strategies

The character of protests is informed by and shapes the geographical contexts in which they occur through at least six spatial strategies, each of which forms the focus of the succeeding chapters.

The strategy of 'know your place', the subject of Chapter 2, refers to how the material conditions of (local) places of work, livelihood and home can generate protests and greatly influence the character of those protests. Collective action by social movements is often informed by cultural practices which are themselves spatially specific, since such practices influence a community's sense of place, that is people's values, beliefs and lifestyles. Chapter 2 shows how farmers in Baliapal, India, drew upon local religious beliefs and practices to inspire and motivate resistance to the construction of a missile base and the eviction from their homes and lands that its construction would entail.

'Know your place' also refers to how activists use their physical and built landscape to shape their protests. The chapter also discusses how, during the 1990 revolution in Nepal, protestors used the rooftops of people's homes to communicate with one another during government curfews, and used community squares as meeting places to discuss tactics because they were out of the reach of government vehicles. Finally, through discussing Palestinian resistance to Israeli occupation, Chapter 2 shows how 'knowing your place' is intimately tied to everyday practices and routines that enable activists to stay put in the face of attempts to displace them.

The second spatial strategy, 'make some space', refers to how protestors actively shape places – by physically transforming the character of, or the meanings associated with, them, for example – as an integral part, or as an outcome of their actions. Through a consideration of anti-roads camps in the UK, the Occupy mobilisations and urban infrastructure 'commons' in Greece, and land occupation in Bangladesh, Chapter 3 discusses how activists use and transform everyday landscapes in the process of protesting, creating not only sites of resistance, but also places where alternative imaginaries and symbolic challenges can be made 'real'.

The third spatial strategy, 'stay mobile', refers to how strategies of mobility – movement in and across space – are critical for the prosecution of protest as well as the necessity of activists to continually adapt to changing contexts and conditions. Through a discussion of the Maoist Naxalite movement in India, the Black Lives Matter and Cop Watch in the United States, and flash mobs and hacktivism, Chapter 4 shows how strategies of mobility generate their own, different geographies. Here, I discuss how particular places may be claimed, defended, strategically used and/or abandoned depending upon the strategies and goals of a particular protest.

The fourth spatial strategy, 'wage wars of words', refers to how activists intervene in and create their own media spaces. It refers to the importance of activist discourses, slogans, websites and social media not only in terms of making demands, but also in creating alternative ways of thinking about particular issues. In Chapter 5 I show how social movements frequently draw upon cultural practices and vernacular languages to inform and inspire collective action while also evoking a sense of place, history and community. Through a discussion of resistance to dam-building in India, the Zapatista struggle in Mexico and various forms of 'culture jamming', I argue that in 'waging wars of words' it is critical for activists to challenge the arguments of the powerful in order to provide alternative understandings of economics, politics and development.

The fifth spatial strategy, 'extend your reach', refers to the communications strategies of protestors and how, as protests around the world become increasingly connected through the use of social media such as, Twitter and Facebook, activists craft and sustain networks of solidarity. Through a discussion of the alter-globalisation network, People's Global Action and activist caravans conducted by the international farmer's network La Via Campesina in South Asia, Chapter 6 shows how social movements mostly operate at the intersection of a series of overlapping scales – from the local to the global – and how different politics of scale provide movements with a range of opportunities and constraints.

The final spatial strategy, 'feel out of place', refers to how particular activist practices challenge everyday assumptions about the meaning and function of places and the type of emotions that are 'appropriate' in particular places. 'Feeling out of place' opens up potentials for people to think, feel and act differently. Through a discussion of the Clandestine Insurgent Rebel Clown Army in the UK, LGBTQ activism and Pussy Riot actions across Europe, Chapter 7 argues that such practices engage with critical emotional and thinking responses in activists, the police and the public that can alter the landscape of protest in important ways.

Sites of Intervention

Drawing upon and extending the original idea of 'points of intervention' created by the smartMeme Collective,[66] I argue that there are at least nine sites of activist intervention that are directly related to a particular protest's concerns, goals or broader strategies.

First, sites of production such as factories, shipyards and fields that reflect activists' struggles to maintain or create sovereignty over the means of livelihood. These can include: offensive struggles, in which workers protest against long working hours and low wages and demand new rights such as rights to organise; defensive struggles, where workers resist the closing down of factories, mines and state-owned enterprises and protest against the restructuring of pay, bonuses and other benefits; and struggles by farmers over land uses. The methods of resistance associated with such sites have included labour and rent strikes, picket lines and occupations (of factories or land), usually at places where such productive activities occur.[67]

A celebrated example of striking took place in Poland during 1980, when 15,000 workers at the Lenin Shipyard in the port of Gdansk went on strike to protest against the communist government's banning of independent trade unions in the country. Activists from the Free Trade Unions of the Baltic downed tools and the strike spread rapidly to other workplaces across the country. The strikers' demands included the right of workers to organise themselves in trade unions independently of state control and the right to strike. Government attempts to halt the strikes – by reforming government-controlled unions, imposing news blackouts and intimidation – only generated more strikes and union organising across different sectors of Poland's workforce. Within a few weeks of negotiations, the government acceded to the strikers' demands and representatives of 35 regions and factories gathered in Gdansk to form the Solidarity trade union.[68]

Second, sites of destruction, such as places of resource extraction (involving such things as deforestation), road and dam-building or military bases that reflect activists' struggles to resist displacement and protest for ecological, economic and social justice. The methods of resistance associated with such locations include efforts at the very locations in which destructive activities are occurring to attempt to prevent or slow down the environmental destruction that accompanies deforestation and the building of roads and dams, or provide public and symbolic critique of destructive practices, such as peace encampments situated at military bases.

A celebrated example of an anti-militarist encampment was that of the Greenham Common Women's Peace Camp in Berkshire, England, from 1981 to 1983. The peace camp was part of the campaign against nuclear weapons that revived in the UK and other parts of the world in the early 1980s in response to the siting of US controlled cruise missiles in Europe. As part of the protests, camps were established at some of the proposed sites, including Comiso in southern Italy, and Greenham Common. These encampments formed part of the opposition to cruise missile sites led by the Campaign for Nuclear Disarmament in the UK. Camps were set up at the access gates to the military base in order to stop the arrival of the missiles. Protestors opposed nuclear weapons in any form, the control of such weapons by a foreign power (the United States) and the use of common land for the establishment of a secret installation from which the public was barred. Protests included 30,000 women encircling the base on 12 December 1981 (the anniversary of the decision to site the missiles in Europe), blocking the roads to the base to prevent vehicles entering or leaving the base, and breaking into the base to obstruct construction activities.[69] Although the camps were eventually disbanded (as cruise missiles were withdrawn as part of international arms agreements between the United States and the USSR), the Faslane Peace Camp continues to this day, opposing the UK government's Trident nuclear submarine base in Scotland.

Third, sites of decision, such as governmental buildings or parliaments and corporate headquarters, that reflect activists' struggles concerning corrupt or unaccountable forms of governance. The methods of resistance associated with such sites involve locating protest at those places where key decisions are made concerning the grievances of protesters. This is because the place of protest has key symbolic value, particularly when covered by the mass media. A particularly potent example of such protests was the march on Washington, DC, in the United States conducted by the US Civil Rights Movement in 1963. Civil rights leaders called for a march and rally in the US capital in order to demand jobs and freedom (from segregation) for African American citizens. After a march from the Washington Monument, approximately 250,000 people assembled in front of the Lincoln Memorial to listen to civil rights leaders, politicians and entertainers. By locating their rally at key sites that represented US democracy and the struggle against slavery, protestors brought their grievances literally to the 'front door' of their government.[70]

Fourth, sites of social reproduction, which refers to the activities, responsibilities and relationships that are directly involved in maintaining protests. The methods of resistance associated with such sites include a range of critical forms of 'everyday activism' that enable and resource social movements, the protests in which they are engaged and the production of activist spaces (such as cooking, childcare and cleaning, as well as sharing stories, facilitating meetings and so on). Sites of social reproduction provide physical and emotional sustenance for activists and operational infrastructures for social movements. Structures of solidarity and social care within protests can be very important. For example, in the various Occupy camps that were set up across Europe and the United States (discussed in Chapter 3), the large numbers of people who came together in the camps required a wide range of care, due to economic recession, the urban location of protests and in some places the weather. Social reproduction thus took the form of cooking for people, dealing with drug addiction and alcoholism, counselling those affected by mental health issues and trauma incurred by police abuse, and providing help for those suffering from exhaustion.[71]

Fifth, sites of circulation such as roads, airports, ports, terminals, squares. The methods of protest associated with these sites can include occupations and blockades of particular infrastructures in order to disrupt the flows of resources, traffic and personnel upon which business and government depend. A celebrated example of such protests was the direct action group Reclaim the Streets, which emerged out of the anti-roads protests in the UK during the late 1990s. Reclaim the Streets critiqued the domination of urban space by the motorcar, which contributed to air and noise pollution and made city streets frequently unsafe for pedestrians. Originating in London, but subsequently spreading throughout the UK and overseas, Reclaim the Streets organised a series of protests that sought to close down roads in cities through staging street parties and celebratory demonstrations. On one occasion, protestors filled a London road with sand and staged a beach party with sound systems, deck chairs and cocktails, effectively closing the road to traffic.

Sixth, sites of consumption, including such spaces as chain stores, supermarkets and art galleries that reflect activists' concerns over the dominant and destructive role of consumerism in contemporary culture. The methods of resistance associated with such sites include such things as consumer boycotts, market campaigns and protests located at those places where consumption occurs. For example, in 2015, the group Liberate Tate conducted a 25 hour protest inside the Turbine Hall of the Tate Modern art gallery in London, urging the institution to drop its sponsorship deal with BP, one of the world's largest oil companies. Seventy-five black-clad protesters covered the Turbine Hall's 500-foot-long floor, which once housed the oil-fired turbines of the Bankside power station, with thousands of words of warning about climate change. The performance's duration represented a full tidal cycle on the River Thames, which is adjacent to the Tate Modern.[72] However, sites of consumption can also be global in character. One of the most celebrated consumer boycotts has been the one against Nestlé, which was generated by health activists' concerns about the consequences of the company's marketing of infant formula milk powder in developing countries. Organised by the Infant Formula Action Coalition from 1977 onwards, the campaign urged consumers worldwide to boycott everyday products produced by the company such as instant coffee and chocolate until the company altered its infant formula marketing practices.

Seventh, sites of potential, which include protests that seek to stimulate the imagination concerning possible future scenarios about how to live and attempt to actualise such alternatives 'on the ground'. For example, Critical Mass are large-scale bicycle rides in which cyclists take over city streets due to the power of their numbers – their critical mass. Begun in San Francisco in 1992, these mass bike rides attempt to bring attention to the second-class status of cyclists on urban streets dominated by the motorcar, and to demonstrate cyclists' right to the road. In addition, the experience of participating in such rides is also important. Critical Mass cyclists are not only occupying urban space through their numbers but opening or transforming that space through generating different experiences and meanings than those inscribed by authorities and made normal through custom.

Eighth, sites of assumption, which attempt to change how people think and feel about particular issues and necessitate challenging underlying beliefs and the control of mythologies. The role of activism here is to hijack events or mass popular spectacles using the images and signs of popular culture. An ongoing example of this kind of protest is the various forms of culture jamming, such as 'subvertising' – bringing together subversion and advertising – referring to making spoofs or parodies of corporate and political advertisements in order to challenge consumers of those advertisements to consider the social or political issues highlighted by the 'subvertisers'. *Adbusters*, a Canadian magazine and a proponent of counter-culture and subvertising, argue that a 'subvert' mimics the look and feel of the targeted advert in order to promote a 'double take' on behalf of the viewer, momentarily challenging taken-for-granted reality. A recent example of such practices was a 'subvert' that mimicked the iPod advert of a black silhouette dancing against a single coloured background while listening to an iPod through white headphones. The subvert showed a black silhouette against the same coloured background but instead of a dancing figure with an iPod, the subvert showed the infamous torture victim image from the Abu Ghraib prison in Iraq: a hooded figure, standing on a box with arms outstretched attached by white wires to electrodes. The word iPod was replaced with iRaq. The 'subvert' was meant as a protest against the US invasion and occupation of Iraq, as well as its torture practices.

Finally, sites of collaboration that engage with certain aspects of the existing hegemony (such as governments) because political critique is never only oppositional, but practised in order to transform the constitutive elements of hegemonic power into new configurations.[73] An engagement with institutions is still possible therefore, but not at the risk of what Antonio Gramsci termed 'hegemony through neutralisation',[74] whereby demands that challenge the hegemonic order are appropriated by the existing system in a way that neutralises their subversive potential. Rather, sites of collaboration are only an effective form of intervention when social, economic or political transformation become possible. For example, struggles for food sovereignty by farmers' movements around the world have, through struggle and subsequent negotiation with governments, succeeded in establishing food sovereignty as part of the national constitutions of 15 countries.[75]

The following chapters will take each spatial strategy in turn and highlight the sites of activist intervention involved. However, as I show in each case study that I discuss, protestors frequently engage with different spatial strategies and sites of intervention at the same time. These spatial strategies and sites of intervention will be drawn together in Chapter 8, the conclusion of the book, in order to map out a radical geography of protest for the future as the planet is faced with a range of economic, political and environmental crises. The chapter will review recent protests in order to show how a radical geographical imagination can lend important insights to both the understanding of struggle and the prosecution of that struggle, opening up new terrains of possibility for activists over the coming decades.

2

Know Your Place

Barricades, Rooftops and Being Steadfast

'Know your place' refers to how the material conditions of (local) places of work, livelihood and home can generate protests and greatly influence their character, and how activists use their physical and built landscape to shape their protests. 'Knowing your place' does not just concern the local physical terrain (although this, as we shall see, can be very important). It is also about knowing the cultural terrain, knowing and being able to draw upon the surrounding culture and its symbols in order to connect with people, motivate them and get them involved in protest activity. The particular cultural, economic and political milieu in which a protest emerges – its place of performance – can influence the character and form of that protest, evoking a sense of place and history. Moreover, the material, symbolic and imaginary character of places can powerfully influence the articulation of protest. For example, a place-specific discourse of dissent can motivate and inform protest (see Chapter 5). This can act as a political disruption and intervention, expressing emotions, hopes and desires, and can give protests their 'feeling space' (see also Chapter 7).

Through a discussion of agitation against the siting of a missile base in Baliapal, India, between 1985 and 1992, the 1990 revolution in Nepal and contemporary Palestinian resistance in the 'occupied territories', I will show how 'knowing your place' enables activists to mobilise from their homes, streets, fields and neighbourhoods and deploy culturally recognisable and powerful symbols that can engage public opinion. The discussion of the Baliapal movement examines how such cultural practices inform protest; the analysis of the Nepalese revolution examines how the built environment was utilised by protestors; and discussion of the Palestinian resistance examines how the materiality of everyday practices can shape protest.

OUR SOIL, OUR EARTH, OUR LAND: BALIAPAL, INDIA[1]

When I arrived in Baliapal, located in the north of the state of Orissa (now named Odisha) on India's Bay of Bengal coast in the autumn of 1989, I was stopped by the police. I was about to cross one of the few approach roads across paddy fields that led to the jungles where the movement was located. A cursory check of my rucksack was made, and after assuring them that I did not have a camera with me I was allowed to continue. After 7 kilometres I arrived at a barricade across the road into the village of Kalipada. The barricades formed the material frontline of a struggle that emerged in 1986 between resident farmers and fisherfolk on the one hand and the central government and military establishment of India on the other.

The government of India had proposed the Baliapal area as the perfect site for a missile test base called the National Testing Range (NTR), to be used for the research, production and launching of missiles and satellites. Baliapal was chosen because of a number of factors: it was away from major towns and ports; it did not affect air and sea routes; it had over 200 days of clear weather per year (being cyclone free); and it was free from oil and mineral deposits. However, the area contained 130 villages of farming families. Taken together with seasonal labourers and fisherfolk, a total of 130,000 people faced displacement if the NTR was built.

In addition, the Baliapal area contains some of the most fertile agricultural land in India, producing a variety of crops that include betel leaf (known locally as *paan*), coconuts, groundnuts, cashews and rice (*padi*) that, together with fishing, had led to the development of a prosperous local economy in an otherwise poor state. While class, caste and gender inequalities certainly existed within the Baliapal area – reflected in the dominance of upper caste and wealthy males within the social structure of the community – most community members benefited from the local economy, and this provided a strong motivation for people to resist the construction of the NTR.

The Baliapal movement was organised by local (dissident) political party activists (symbolising representational power) and was led by an assortment of groups. There was an all-male steering committee that made decisions regarding movement strategies and tactics, a council of landless peasants, fisherfolk and women, and four activist fronts (one each for students, women, youth and fisherfolk) that integrated the class, caste, gender and political divisions that existed within local society. The movement was able to mobilise almost 50,000 farmers. Protests emerged in sites of production (the fields and rivers where people made their livelihoods) and sites of social reproduction (the homes and everyday practices of farmers and fisherfolk), and targeted the potential site of destruction (of farmers' livelihoods and culture) represented by the planned missile base.

Our Soil

Women were particularly important in the movement since they did much of the agricultural cultivation and were intimately tied to the natural environment, their militancy being motivated by their perceived need to protect their right to the soil and to continue their lineage. Indeed, the Baliapal movement was informed and motivated by a potent sense and knowledge of place that refined and strengthened the economic motivation provided by the local economy. This was epitomised by the movement's ideology of *bheeta maati* ('our soil'), articulated as 'our soil, our earth, our land'. As one farmer remarked to me: 'For Baliapalis the land is our mother, our earth, our home. This is in the hearts of the people'.

Through the articulation of *bheeta maati*, Baliapali women drew upon the Hindu and peasant folk culture of which they were part. In Hindu tradition, the earth is a purifying agent and remedy for diseases, and acts as a provider of strength and (agricultural) abundance. Throughout much of India, the land is worshipped as 'mother earth'. In Orissa, every village territory has a goddess who is an incarnation of Kali, the mother goddess of time, creation, destruction and power. Those born on the same portion of earth, in the same village, share the same mother, namely the village goddess. As her children, they are all one and form a kin-like community, the goddess being the earth of the community.[2]

The movement's ideology drew upon the cultural and economic dimensions of the peasants' everyday life. It converted the cultural sentiment for the land – the peasants' sense and knowledge of place – into a political demand for the absolute right to its continued use through religious and mythical tools. Stories from the *puranas* (sacred Hindu texts) and the *Mahabharata* (one of the most famous Hindu epics) were intertwined with political action to confront the state. Movement ideology was articulated in songs and dramas that emerged as a potent expression of the Baliapal movement's war of words against the government. As activists informed me, the songs and dramas were performed locally in Baliapal's villages to empower and motivate people to continue the struggle, to provide an ongoing oral history of the movement, to inform people about current developments in the struggle, to teach non-violent methods of resistance, and to increase awareness regarding the agitation.

Barricading the Villages

The movement's compositional power made some space by constructing barricades at key sites of circulation, erected across the four approach roads that crossed the paddy fields that surrounded the 130 villages. These were constructed to prevent the entry of government officials. Blockades are both material (they seek to achieve a specific quantifiable objective) and communicative (they communicate a political position, a set of values or a worldview).[3] The barricades were targeted at the state and its plans for a missile base, and were also symbolic acts undertaken to amplify messages of resistance. On the barricade at Kalipada, a sign in Oriya stated: 'Land is ours, sea is ours. Government officials go back'. Upon the gates, written in English, were the words: 'We want land, not missile base. We want peace, not war'. Such communicative action can contain powerful expressive outcomes by building the resolve and commitment of participants (see also Chapter 5).

In order to warn villagers of approaching government vehicles, the people staffing the barricades beat metal plates (*thali*) and blew conch

Figure 2.1 Know your place: women blow conch shells to alert villagers, Baliapal, India, 1989. Photograph by the author.

shells – found in every Hindu home in Baliapal, being used to summon the important local deity Jagannath (Lord of the Universe, whose major temple is located in Puri, Orissa). When the conch shells were blown, villagers left their work and congregated at the barricades, whereupon they lay down in the road forming human roadblocks to prevent entry by the government.

An example of the people's militancy was illustrated when, in February 1988, 24 magistrates, accompanied by 3,000 armed police attempted to enter the Baliapal area in order to 'explain' to the residents the government's plans for the villagers. They were confronted by 20,000 women, children and men forming a human blockade and were prevented from entering the area. Villagers stayed mobile, responding to changing circumstances, as one farmer told me: 'People work as usual, but if the call comes, then we drop whatever we are doing and rush to the barricades. We don't believe in the government, we believe in people's power'.

The Maran Sena

The movement also established a 5,000 strong *maran sena* ('suicide squad'), of whom 1,500 were women. The *mara sena* pledged in blood before the village goddess Durga (an incarnation of Kali) to give their lives by throwing themselves under the wheels of approaching government vehicles rather than allow the barricades to be breached. What was particularly potent about the *maran sena* was that it drew upon the local community's knowledge and connection to place. Once a year in Puri, the Rath Jatra festival celebrates the key Hindu deity of the region, Jagannath, by wheeling a giant chariot around the deity's temple. Over the years, onlookers have fallen under the wheels of the chariot and been crushed to death. In Orissa this is considered a sacred death – the deceased is transported directly into the arms of Jagannath. The *maran sena* directly drew upon this religious symbolism in threatening to throw themselves under the wheels of government vehicles. By pledging before the goddess Durga, the movement's war of words communicated to the government that they were deadly serious in their intent to prevent officials from entering the area. The *maran sena* slogan was, 'After killing me the Range will be established on my corpse'.

Self-Government

Finally, the movement drew upon traditional local practices and the community's relational power in order to manage disputes in the area. This was facilitated by the revival of the *vichar* institution, whereby discussions concerning village problems (such as land disputes) found solution in a consensus process within village councils without recourse to the law enforcement and judicial establishment. The institution was an effective instrument in mediating conflicts over both land and property cases. As one activist informed me, between 1985 and 1990 over 400 land disputes were settled, some of which had been pending for the past 30 years. In addition, this process was supported by the refusal of villagers to pay back loans or pay their taxes. Moreover, the movement sought to extend their reach by eliciting support from local media and state officials of opposition political parties (as a site of collaboration).

The state government responded to the movement by introducing an economic blockade of the area. They arrested and detained activists and deployed 8,000 armed police around the area. However, the movement stood strong. Knowing their place was a source of profound strength and inspiration to the protestors. For example, during December 1986, 45 armed personnel of the Central Reserve Police Force attempted to enter the area. As one farmer told me:

> The police were armed with halogen lamps, water cannons, tear gas and petrol to ignite the villages. As they attempted to enter the area they were met by hundreds of women who said: 'Sons, you have mothers like us. If you are ready to kill your mother, then fire at our hearts'. The women were followed by children, of ten to fifteen years old, who addressed the police: 'Father, you have sons like us. If you kill our parents, how will we live? If you kill our parents, then take us with you'. At this the children began to cry. Incredibly, the police also began to cry and lower their guns. They embraced the children as the local people began to applaud and shout, 'The police are our brothers'.

In February 1990, elections in Orissa saw the political party Janata Dal elected to power in the state assembly. They had opposed the NTR while in opposition and deployed their representational power to drop plans for it to be located in Baliapal. Five years of resistance drawing upon peasants' deep knowledge of their place of livelihood had ensured the continuation of the livelihoods and culture of the people of Baliapal. When I visited Baliapal in 1992, I was told that the movement was sleeping, but ready to mobilise again if necessary. While Baliapalis constructed barricades to defend their homes and lands, in Nepal protestors used their knowledge of urban topography to prosecute revolution.

ROOFTOPS, SQUARES AND BACKSTREETS: KATHMANDU, NEPAL[4]

From my room in Durbar Square in Kathmandu's old city, I could see right across Nepal's capital. At a pre-arranged time, the lights across the city went out. On the streets, protestors confronted armed security forces. Amidst tear gas and bullets, protestors' fires burned, activists' shadows casting flickering arabesques against the buildings of the old city. The protests had commenced a few weeks earlier in February 1990 as a popular movement sought to transform the country's political structure. A poor, predominantly rural country, Nepalese society was rent with social, economic and political inequalities compounded by a caste system dominated by a small minority of high-caste Brahmins, Chhetris and Newars, who dominated economic and political life in the country. The majority of the lower-caste population were condemned to the ravages of poverty, bonded labour, illiteracy and inadequate health care. In 1990, over 60 per cent of Nepal's population of approximately 18.5 million lived beneath the official poverty line.[5]

In contrast, the royal family, headed by the (Chhetri) king, controlled the social order in alliance with a religious hierarchy, the army, landed interests and modernising capitalist, bureaucratic and political elites. This control was maintained through the coercive force of the army, controlled by the king, and through inequalities of power institutionalised in the *panchayat* system. Based upon a pyramidal structure of village, district, zonal and national councils, with the king at the apex, the *panchayat* system provided the king with autocratic power over the political, economic and cultural life of the country. This power was institutionalised through a variety of repressive laws, designed to curtail any opposition to the government. Both parliament and the judiciary answered only to the king; freedom of peaceful assembly and association was denied (indeed, political parties had been banned for 30 years), imprisonment without trial was legally sanctioned, and torture and deaths within prison were commonplace.[6] The Movement for the Restoration of Democracy (MRD) emerged within this context, determined to transform Nepal's polity.

Emergence

Inspired by the success of popular movements in Eastern Europe in 1989, Nepal's principal opposition parties, the Nepali Congress and the

United Left Front – the latter a coalition of seven of Nepal's communist parties – decided, despite ideological differences, that it was in their mutual interest to join together to launch the non-violent MRD. The organisational power of the movement targeted the representational power of the state. Their demands focused on the dismantling of the *panchayat* system, the restoration of parliamentary democracy and the reduction of the king's powers to those of a constitutional monarch. In so doing, protestors intervened in sites of assumption, challenging the continuance of the status quo, and sites of potential, proposing a more democratic political system.

During the revolution, outbreaks of resistance were widespread throughout Nepal, as the MRD extended their reach beyond Kathmandu through demonstrations in the urban centres of Pokhara, Biratnagar and Janakpur as well as other towns in the southern Terai region of the country. However, the principal terrains of resistance against the regime were located within the Kathmandu Valley, particularly in the capital city of Kathmandu and the surrounding towns of Patan, Kirtipur and Bhaktapur. Kathmandu and Patan in particular emerged as sites where the power of the royal regime was contested. Their urban landscapes not only influenced the articulation of resistance, they were also transformed into a tool of struggle as activists utilised their intimate knowledge of the city's urban topography.

Kathmandu's location as the capital of Nepal, the home of the king (and hence the centre of royal power) and the locus of representational, administrative and economic power made it a compelling site of decision for the contestation of such power by the MRD. This was further facilitated by the improved communications that existed within the city and the valley of Kathmandu, the fact that communications outside the valley were difficult and that there was a dense concentration of population within the city (400,000 people) and the valley (800,000 people) at the time. These factors facilitated mass mobilisations within urban areas that were not possible elsewhere in the country. Kathmandu was also the centre of organisational power of the MRD, as it was the location of the principal headquarters of the main opposition political parties and of the student organisations involved in the movement.

Rooftops and Blackouts

The MRD deployed a repertoire of protest tactics, such as labour strikes (targeting sites of production) and demonstrations, as well as those that drew upon activists' place-specific knowledge. For example, an important tactic utilised during the revolution was the blackout protest, whereby all of the households in Kathmandu and Patan were asked to turn out all their lights as a symbol of dissatisfaction with and resistance against the government. These protests were often called during the evening curfews that were imposed by the government in an attempt to quell the movement.

Politically, black represents the colour of refusal and dissatisfaction in Nepal. In blacking out the city, residents symbolically communicated mass resistance to the royal regime, forming part of the MRD's war of words, its discourse of dissent. The blackouts also enabled city residents to grasp the extent of popular support for the MRD, and acted as a morale booster to the movement. They also enabled increasing numbers of people to show solidarity with the movement, and to challenge the curfew and join demonstrations under cover of darkness with a reduced chance of being identified by security forces.

Although the blackouts were organised by the movement's leadership, it was the cities' residents who communicated them. Residents relayed the message of the action from rooftop to rooftop across Kathmandu. In doing so, they drew upon and utilised their knowledge of everyday practices particular to the place. Traditional Newar houses within the city consist of three, four or five storeys. The upper storey opens out onto a porch (kaisi) that is used for various parts of rituals. One of these – the flying of kites during the Mohani festival as a message to the deities to bring the monsoons to an end – involves symbolic communication.[7] The porches are also sites of social reproduction used for everyday activities such as the drying of clothes and chilli peppers, and talking with neighbours.

By informing their neighbours of the blackout protests from their rooftop porches, residents deployed their relational power and utilised a cultural space that was already important for both community and symbolic communication. In so doing, a space of interwoven meanings was produced. The rooftops acted as a place for the performance or religious rituals and daily activities, and facilitated strategic communication of the resistance.[8] The latter was facilitated by the propinquity and low elevation of city dwellings, and the fact that they were out of the purview of government forces. Once on the streets, people targeted sites of circulation by setting fire to car tyres to act as temporary barricades across the narrow streets. Pitched battles between armed riot police and stone-throwing demonstrators ensued, the incendiaries of protest lighting up the darkened city.

The public contestation of space transformed the physical character of Kathmandu and Patan, inscribing upon them a mosaic of signs that spoke of the ferment in their streets. The MRD's war of words continued with photograph displays erected in public squares depicting victims of police torture and killings, political prisoners and activists who had disappeared. Numerous windows of government offices and shops were broken. In the middle of roads, burned-out skeletons of government buses attracted crowds of onlookers. Torn-up street stones, used in battles with the police, lay strewn across streets and pavements. Upon city buildings and temple walls were daubed pro-democracy and political party slogans.

The Backstreets and Squares of Patan

The town of Patan, close to the capital, served as the base of operations for the movement's underground leadership during the duration of the MRD. The movement consciously used their knowledge of the urban topography of Patan to accommodate the exigencies of underground activism. Within Newar towns, community relations are focused around community spaces called *twa:*, which represent important loci of personal and household identification. Usually, a *twa:* will be centred around a spacious square at the centre of a matrix of narrow winding streets and bazaars. These squares are usually paved and used for various agricultural and commercial purposes, as well as serving the immediate communities as a focal point where the inhabitants of that particular community meet.[9]

The numerous squares within Patan were used by the MRD as meeting places for the discussion and planning of movement strategies. Being interlinked within a labyrinthine web of streets, the squares afforded a protected space out of the purview of the government. The narrow streets of the town prevented any mass deployment of government forces, or the deployment of armed vehicles, while aiding the escape of activists from the police. The interconnected network of backstreets that traversed the town enabled activists to avoid the main streets and to stay mobile, moving unhindered from one end of Patan to the other, and from Patan to Kathmandu without detection. Throughout the revolution, the movement was able to consciously utilise their knowledge of urban space to break government curfews, and maintain communications between movement members in Patan and Kathmandu concerning actions and conduct meetings. Indeed, this use of urban space enabled the underground leadership of the MRD to avoid capture during the entire uprising.

The most dramatic use of urban space occurred when Patan was defended against incursion by government forces for a period of one week during the revolution. During a blackout protest, deploying compositional power, MRD activists made some space at sites of circulation by barricading the seven approach roads to the town and digging trenches to prevent the entry of government vehicles and personnel. The effectiveness of the barricades was enhanced by the fact that there is only one major entrance to Patan from Kathmandu, across the Bagmati River via the Patan Gate. When I arrived at this gate on the second day of liberation, two lines of peasant farmer women, armed with their agricultural tools, staffed the barricade across the road. Smoke from burning tyres spiralled into the sky behind them.

The movement drew again upon local everyday practices and spatial knowledge to defend Patan from incursion by the government. In Newar culture, social relations are arranged in successively inclusive units – the household, the *twa:*, the town and so on – whose boundaries must be protected at each level. Because it is the focus of face-to-face communities beyond the extended family, the *twa:* is the realm in which individual and family reputation is at risk. To ensure moral integration within the wider community, individual and household reputation must be protected from (moral) incursion. Within the town, the *twa:* is a significant locus of action in itself (during household religious ceremonies and funeral processions, for example), and in concert with the other *twa:* that constitute the town (such as during various religious festivals). Because of their social significance, the reputation and moral integration of *twa:* (and thus of their inhabitants) within the wider community of the town was also to be protected.[10]

A member from every household in the *twa:* nearest each barricade volunteered for the security committees to staff the barricades. In defending the town from incursion by the government, each member of the committee was also protecting their *twa:* and their household. The traditional importance of ensuring moral protection within the boundaries of household, *twa:* and town was transcribed onto movement organisation for the defence of Patan. The security committees consisted of between 50 and 100 people, who were armed with knives (*khukri*), tools and sticks, and staffed the barricades around the clock. All those entering the town were checked. The MRD stayed mobile: to warn of the approach of government forces, temple bells were rung, whereupon people would rush to the barricades to prevent the government's entry.

At night, wood and tyres were burned in the streets and residents would gather en masse to protect the town and conduct torchlight demonstrations. In addition, activists surrounded the police station at Mangal Bazaar in the heart of Patan. I was told by activists that the 128 police at the station had been informed that if they remained inside, protestors would bring them food and water. However, if the police attempted to leave the station, activists warned them that they could not guarantee their safety. During its liberation, the meanings associated with Patan as a site of social reproduction – a place of home, community and Newar culture – became interwoven with those of Patan as a place of activist networks, meeting sites and organisation. For the activists whom I spoke to, Patan acted as a material and symbolic site of resistance to the royal regime, one that served to inspire the imaginations of the residents of the Kathmandu Valley during their protests against the government.

After one week of liberation, the army intervened in Patan. Movement leaders told me that they decided not to resist the army because they wanted to avoid direct confrontation with the military that would cause unnecessary bloodshed, and they did not want to delegitimise the MRD's predominantly non-violent character. Despite government intervention in Patan, the town continued to act as the base of underground operations of the movement.

Human rights and student groups within the MRD made use of a well developed underground press to wage a war of words against the government, disseminating information concerning movement activities and the government's human rights abuses. Photocopying machines were used to print daily action reports, in Nepali and English, that were circulated both inside and outside Nepal. Events taking place within the urban spaces of the Kathmandu Valley were relayed to other places of struggle within Nepal, serving to inform and inspire those resistances and extending the reach of the MRD. The availability of fax technology enabled the movement to communicate events to six international sites (including the United States, Germany, Australia and Switzerland) and keep the world watching, reflecting the conscious location of movement action within global (as well as local and national) space. In so doing, the movement was able to elicit international support from US, German and Swiss aid agencies, who threatened to withdraw aid to the Nepalese government if the repression of the movement continued.

The government's response to the movement was predominantly coercive, ranging from arrests and detention to torture and police shootings of activists. Massive deployments of police and security forces were made in Kathmandu, and police observation posts were erected throughout the city. Armed police set up roadblocks in parts of the city and along the roads leading out of the capital. Reports of police shootings in Kathmandu, Bhaktapur, Patan, Kirtipur and elsewhere in Nepal were widespread.[11] The most notorious example occurred during a countrywide strike called by the movement on 6 April 1990 when 200,000 people held a non-violent demonstration against the king outside the primary site of decision, the royal palace in Kathmandu. Following an attack upon the crowd by police armed with *lathis* (wooden sticks), some demonstrators responded by throwing bricks at the police. The security forces pumped tear gas into the crowds and then, as demonstrators broke shop windows and defaced a royal statue, the security forces opened fire on the crowd, killing at least 50 people (although human rights groups put the number closer to 500) and injuring hundreds of others in what became known as the 'Massacre of Kathmandu'.

Two days after the shootings, the king announced that the ban on political parties had been lifted. The MRD intervened in a site of collaboration as consultations between party political leaders and the king culminated in the movement being called off. An interim government, including representatives of the Nepali Congress and United Left Front, was established. Across Kathmandu, the flags of previously banned political parties appeared, hung from rooftops and displayed in shop windows. Police observation posts were dismantled and constant police patrols of the city were curtailed. By mid May, a constitution committee was formed by the interim government, paving the way for parliamentary elections the following year.

Despite this success, many of the fundamental inequalities that permeated Nepalese society remained intact. The king retained control of the army, and the range of ongoing economic, caste, ethnic and gender inequalities that permeated Nepalese society were left largely unchanged. The unfinished character of the revolution was to ultimately lead to a ten-year Maoist insurgency (between 1996 and 2006) and another popular revolution in 2006. This culmination of this was the events of 22

April 2006, when hundreds of thousands of people filled Kathmandu's 27 kilometre-long ring road, effectively encircling the city, sealing off one of the most important sites of circulation in the city. After 19 days of popular uprising across the country, the Nepalese monarchy was abolished and the country became a republic. Nevertheless, social, political and economic inequalities persist, and resistance amongst various ethnic groups has continued in the south of the country.[12] While the Nepal protests were an example of transforming the use of everyday spaces, in Palestine, everyday activities can themselves constitute resistance.

BEING STEADFAST IN PALESTINE

The history of Jewish settlement in Palestine dates back to the nineteenth century, although it significantly increased during the British Mandate for Palestine in the 1920s. By the time of the establishment of the state of Israel in 1948, Jews, who comprised 35 per cent of the population, controlled 56 per cent of the territory. The seizure of Arab cities and destruction of Arab villages and neighbourhoods intensified during what Palestinian's refer to as the Nakba or 'catastrophe' – the destruction, dispossession and displacement that followed Israeli nation-building and land annexation. By 1949, 750,000 Palestinians – half of the Arab population – had been displaced. Between 1948 and the present, Israel has continued to destroy Palestinian villages and build Israeli settlements on land illegally confiscated from its former Palestinian occupants. Arabic place names have been replaced with Hebrew ones. Although the Gaza Strip is under Palestinian control, the West Bank area is being steadily occupied by Israeli settlements. The process of land appropriation and illegal settlement continues, accompanied by the violent repression of Palestinian resistance, including killings, arrests, detention, curfews, border closures, roadblocks and house demolitions.[13]

Knowing your place is central to Palestinian resistance to Israeli occupation. The particular characteristics of place are used as a tool to reinforce and support ongoing resistance to occupation. Asef Bayat has insisted on the importance of everyday life and routine practices in Middle Eastern politics, through what he terms the 'art of presence':

> active citizenship, a sustained presence of individuals, groups and movements in every available social space, whether institutional or informal, in which it asserts its rights and fulfils its responsibilities.

For it is precisely in such spaces that alternative discourses, practices and politics are produced.[14]

In the context of Israeli–Palestinian politics, the ownership and control of space are central issues. Physical presence can enable the control of space, providing an ability to act within that space, and to materially and symbolically contend with the dominating representational power of the Israeli state. The 'Palestinian resistance' is fragmented into multiple organisations – many of which have their own armed wings – and further supplemented by independent popular resistance committees. These various groupings shift between cooperation, competition and conflict.[15]

For Palestinians, human presence is marked physically through signs that name places, since the naming of places can itself be a challenge to official (Israeli) accounts of territory that wish to efface Palestinian presence and claims to land. This is particularly true of property rights, since Palestinians are faced by a legal apparatus that justifies land seizure from them and the construction of illegal Israeli settlements on their land. For Palestinians, continued presence in places (villages, towns and so on) are in effect a particular spatial strategy employed to defend their homes and lands. Presence represents attachment to their land and demonstrates the legitimacy of Palestinian claims to the land. It also acts as a deterrent to Israeli settlers or soldiers who might attempt to appropriate Palestinians' homes if they are empty.[16]

Place-naming compliments physical presence, in that it affirms the existence of a place and symbolises its particularity, and shows opposition to the Israeli state. This is important because of the erasure of Palestinian names following the destruction of Palestinian villages and neighbour-hoods. Palestinian activists deploy compositional power and make some space by naming buildings, streets, neighbourhoods and villages in Arabic, either through signs that mimic the official ones or through graffiti, in order to publicly declare their attachment to particular places, to affirm cultural memory and to show that Palestinians know their places.[17] Such a spatial strategy is also at the same time part of a broader war of words that is waged by Palestinians in their struggle for nationhood.

Palestinians' practice of physical presence in particular places is known as *sumud*. *Sumud* designates steadfastness, a stubborn presence in place and unwavering commitment to Palestinian land. In particular, women's (gendered) role as guardians of the home means that through their everyday life practices they intervene in sites of social reproduction such as participating in community events and festivals.[18] Staying in one's homes and villages becomes the very act of resistance in the face of erasure. It means holding onto the place, not succumbing to intimidation by the Israeli authorities and not abandoning houses that might otherwise be taken over by settlers. *Sumud* also means inhabiting a home, living in a neighbourhood despite ongoing hardships and going through the practices and routines of everyday life. Inhabiting the home means resisting pressure by the Israeli government, army, police and settlers through orders of destruction, actual demolitions, restrictions on building permits, invasion of property or through attempts to buy or occupy Palestinian property.[19]

In addition, protection of domestic space is facilitated by active work in the neighbourhood and community to improve the quality of life and interpersonal bonds (strengthening relational power) in order to make communities more resilient and make *sumud* more successful. Compositional power is deployed through numerous community-building initiatives to make some space such as neighbourhood centres, health care centres, recreational and meeting spaces. They not only enable the strengthening of community bonds, but also provide protection for children from arrest and military violence, and from the general atmosphere of oppression that accompanies occupation. The practices of everyday life can thus facilitate continued Palestinian existence in place, generating feelings of being at home, being part of a community and, over time, of some form of permanence of inhabitation in a context of occupation.[20]

KNOW YOUR PLACE

Protestors consciously utilise knowledge of their local home places as a spatial strategy to prosecute resistance. In the villages and jungles of Baliapal, the streets and squares of Kathmandu and in Palestinian homes, the ability to 'know your place' enabled protestors to deploy powerful, culturally recognisable symbols in order to engage public opinion and participation. The media has a tendency to 'flatten' most expressions of protest, focusing upon the more spectacular aspects of a given confrontation (such as a mass demonstration filling the streets of a capital city or a labour strike inconveniencing commuters) while tending to render similar very different forms of protest. However, as I have shown, protestors frequently use place-specific knowledge and cultural codes.

These examples of 'knowing your place' also attest that while political action is frequently informed by local conditions, it is also the product of wider sets of relations, processes and connections. For example, resistance in Baliapal was a response to political decisions made in Delhi, the Indian capital, concerning the location of the NTR, and developed solidarities with political parties at the Orissa state level. The Nepalese revolution – while focused on Kathmandu – was national in character and communicated internationally through various media. And the Palestinian struggle has been national and international in character since its inception, contributing to geopolitical processes across the Middle East and beyond. Nevertheless, as I have shown, the act of protest also partially transforms the spaces in which it takes place – be it through the construction of barricades, the liberation of towns or the construction of community centres. In this sense, protest is always in some ways an act of making space as well, and it is to this that the next chapter is devoted.

3

Make Some Space

Camps, Commons and Occupations

'Make some space' concerns how protestors physically transform the character of or meanings associated with places as an integral part or outcome of their actions. The transformation of places can form a deliberate strategy to further protestors' campaign goals, and that protest activity can imbue places with the cultural meanings, memories and identities of protestors. Activists use and transform everyday landscapes in the process of protesting, creating not only sites of resistance but also places where alternative imaginaries and symbolic challenges can be made 'real'. The act of 'taking space'[1] also necessitates practices of fixity (that is, staying put in a place, such as the Palestinian resistance discussed in Chapter 2) and mobility (which I will discuss in detail in Chapter 4). 'Making some space' is a critical component of protesters' toolboxes since the act of transformation conveys potent symbolic messages and images of alternative ways of doing things to society at large.

For Raul Zibechi, the art of making space in community-based struggles potentially enables a dispersal of power from the state and capital.[2] The social relations generated comprise spaces 'in which to build a new social organisation collectively, where new subjects take shape and materially and symbolically appropriate their space.'[3] For Zibechi, territory is *the* crucial space in which contentious politics is fashioned, understood as both material territory (involving struggles over the access, control, use and configuration of environmental resources such as land, soil, water and biodiversity, as well as the physical territory of communities, infrastructure and so on) and immaterial territory (involving struggles over such things as ideas, knowledge, beliefs and conceptions of the world).[4] For indigenous people, territorial struggles comprise resisting the theft of their land (and other resources) and the appropriation (by capital and the state) of indigenous sovereignty and indigenous peoples' relationship to their land and communities.[5]

Such territorial struggles are conceptualised by Zibechi as 'societies in movement', defined through their creation of social relations of autonomy characterised by the (re)appropriation of resources, increased potentials for cooperation and transversal connection, the generation of new types of knowledge and capacities that facilitate self-organisation and more horizontalist (that is, non-hierarchical) organisational forms.[6] Many of these characteristics have been termed forms of prefigurative politics (that is, living in the present the future that is desired). The spaces they make represent 'commons' – resources that are collectively owned or shared between people – and 'the common', that is the relational power generated by people acting and being-in-common.[7]

'Making some space' is also about articulating the symbolic significance of particular spaces and the protests that take place within them. Practices of prefigurative politics, or 'societies in movement', symbolise disruption of and resistance to the status quo of neoliberal capital accumulation and offer alternatives to it.

In order to draw out these dynamics I will consider a range of different examples. First, I consider an anti-roads protest in Glasgow as an example of material and symbolic resistance by environmental activists, manifested in a temporary encampment that drew upon creative compositional power. Second, I consider an example of territorial struggle in Bangladesh. In contrast to the first example, the intention behind the occupation of land – as an example of a society in movement – was to secure permanent livelihoods for poor landless peasants. Third, I consider the occupation of Tahrir Square in Cairo, Egypt, the Occupy mobilisations and subsequent alternative urban infrastructure 'commons' in Greece. All of these provide examples of compositional power, articulate different forms of systemic critique (of military rule, economic inequality generated by neoliberal capitalism and austerity policies respectively) and represent very different forms (and outcomes) of societies in movement.

POLLOK FREE STATE[8]

Amid the sounds of car horns, whistles and cheers from the assembled crowd, cars driven to Pollok Free State by Greenpeace activists were manoeuvred onto the road where tombs have been dug for them. Engine down, and with earth and stones packed around them, we buried the cars vertically in the road. A great cheer rose from the crowd as one teenager from the nearby Pollok housing estate hurled a stone through the driver's window of a car, shattering the glass. A resident of the Free State swung a sledgehammer and dispensed with the windscreen. Another cheer rose from the crowd. We were a rhythmic crowd, moving to the visceral beat of drums. We revelled in the burial of the car, encoded as it was with our resistance to the construction of the M77 motorway. Once the cars were buried, petrol was poured over them and they were set alight. The cars were then spray painted with political slogans. Voices of celebration filled the air. People danced in the firelight, which cast shadows of celebration upon the road. We danced fire. We became fire. Our movements became those of flames.

Pollok Free State was a protest encampment set up as part of mobilisations against the M77 motorway in Glasgow, Scotland. These in turn formed part of a broader wave of anti-roads protests that emerged in the UK during the 1990s, challenging the government's road-building programme. The M77 motorway – an extension to the A77, the main road from Glasgow to Ayr – was planned so that it would run through the western wing of Pollok Estate – an area of farm, park and woodland stretching over 1,118 acres, 4 miles south of Glasgow's city centre.

While Strathclyde Regional Council (together with the Scottish Office) argued that the road would relieve traffic congestion, opponents' war of words argued that the road would cause increased noise and air pollution from increased car use, cause irreparable damage to the woodland and wildlife habitats of the western wing of Pollok Estate, generate increased traffic and divert resources that could be used to upgrade existing transport facilities. The construction of the road would also restrict the access of local low-income communities to Pollok Estate – which is a safe recreational area for children – and place a loud, polluting motorway close to primary and secondary schools.

In April 1994, the Stop the Ayr Road Route (STARR) Alliance was launched, merging the organisational power of a range of community and environmental groups resisting construction of the M77, including Friends of the Earth (Scotland), the Scottish Wildlife Trust, Earth First!, Corkerhill Community Council and Greenpeace. The focus of resistance was the protest camp, Pollok Free State.

Pollok Free State was an encampment of tree houses, tents and benders (do-it-yourself dwellings) located at a site of destruction and circulation in Barrhead Woods of Pollok Estate in the path of the projected motorway. Located south of Glasgow's River Clyde, the Free

State was located close to several low-income housing estates, including those of Pollok, Corkerhill and Arden. Established in June 1994 by Colin McLeod, an Earth First! activist and a Pollok resident, the camp acted as a visible symbol of resistance to the motorway.

As an act of compositional power, the Free State symbolically announced the act of claiming, bordering and making space, as well as the contest over land use.[9] The making of space in protest camps involves various objectives and operations. These include establishing media and communications infrastructures that provide such things as mainstream media tents, media liaison, phone trees and activist media; and establishing everyday protest activities that constitute protest camps as sites of social reproduction. The latter include legal, medical and activist trauma support, governance infrastructures such as meeting spaces, announcement boards, decision-making policy guidelines and so on, and activities that reproduce everyday life in the camps such as dealing with food supply, cooking, shelter, sanitation and the maintenance of communal and private space.[10] Where police liaison is necessary, protest camps also become sites of collaboration.

Pollok Free State represented the 'homeplace' and the focus of resistance against the M77, articulating an alternative space that occupied symbolic and material locations. It acted as a place where people who were interested in the M77 campaign could learn more and get involved. As a site of potential, the Free State stood as a critique of the environmental damage caused by road-building and an example of how people might live their lives differently, issuing its own passports as symbols of such potential.

As such, the Free State represented an experiment in do-it-yourself ecological living, using alternative technology (such as wind-powered generators) and constructions that were out of place in a typical wooded area such as tree houses, benders and eco-art (carved wooden totems of owls and ravens, for example). The Free State developed relational power, attracting a variety people from the Glasgow area – including artists, scaffolders, tree surgeons, carpenters, musicians and cooks – who contributed their skills to the camp. In addition, people from the surrounding housing estates visited the camp to participate in ongoing work, donate food and so forth.

Pollok Free State was a site of assumption, challenging the idea that further road-building was the most appropriate transport planning solution, and it waged a war of words against transport planning policy, as can be seen in the 'Declaration of Independence' that was printed on the Pollok Free State passports:

> Our land is threatened with destruction in the name of 'infrastructure improvement'. Pollok Estate ... is threatened by privatisation for a car owning elite. ... These actions have been taken without consultation and imposed upon us. We therefore maintain that the threat to our environment and liberty by this road and legislation is incompatible with sustainable environmental use and any notion of democracy. We the inhabitants of Pollok Free State ... call on all people who share these beliefs, ideals, and aspirations to come to the defence of this new domain.

Activist strategies included staying mobile, whereby small clandestine groups of so-called Pixie Patrols would also attempt to delay and disrupt tree-felling activities by physically disabling bulldozers. Activists also attached themselves to bulldozers, chainsaws and security vehicles with kryptonite bicycle locks (an activity known as 'locking on') so that the machinery could not be used until they were removed, and occupied trees (in tree houses, for example) to prevent their felling.

While direct action was effected in an attempt to prevent the construction of the M77, it also imposed upon the road-builders extra security costs and delays. As disruptive direct action increased, so the road constructor Wimpey was forced to hire increased numbers of security guards to protect their equipment, and to enable their staff to proceed with construction of the road. In this sense, direct action was a 'market force'.

However, the most visually dramatic symbol of the campaign was the creation of Carhenge, reflecting the compositional power of the protestors. This consisted of nine cars (eight in a circle with the ninth in the middle), buried front-end-down in the surface of the M77.

This site of assumption not only humorously evoked Stonehenge; it was also a symbol of the end of the age of the car, a dark work of art which challenged people's common-sense assumptions about car culture. Carhenge was symbolic of what activists understood as the system's irrationality – the poisoning of the air that we breathe by increasing amounts of exhaust fumes, or 'carmageddon'. It was also a challenge, activists declaring that the nine buried cars would be the only vehicles that would use the M77.

Figure 3.1 Make some space: Carhenge, Pollok Free State, Glasgow, 1994. Photograph by the author.

As a site of potential, Pollok Free State represented a life with fewer controls, as one activist, highlighting relational and organisational power, commented:

> You know, these months [in the Free State] were more important than stopping the road. Life is more real ... I felt free for the first time. For the first time I lived life without constraints imposed by the government. People were living, working together, a community, showing that there are those who resist.

The protest also extended its reach through the use of various media, including a report by *Undercurrents*, the activist news video organization, along with those of local radio, television and national newspapers, and made connections across the UK with other anti-roads protests. Pollok

Free State attempted to retransmit to the system its own contradictions particularly with regard to the government's road-building programme. In doing so, the Free State created a temporary public space wherein its residents could represent their counter-cultural, environmental views.

Although the protesters were able to delay the felling of trees around Pollok Estate, eventually the construction of the M77 proceeded. However, most of Pollok Free State and 50 trees around it were defended successfully, although the Free State was subsequently abandoned in early January 1996. Anti-roads protests continued to spread across the UK during the 1990s and also gave rise to the activist group Reclaim the Streets, who oppose the car as the dominant mode of transport, the privatisation of public space and the neoliberal economics that underpin it. While Pollok Free State provides an example of the creation of temporary resistance encampments to prosecute protest, in Bangladesh land occupations attempt to make some space more permanently.

LAND OCCUPATION IN BANGLADESH[11]

'The sky has gone to bed', commented a Bangladeshi friend as we trundled by rickshaw down the dirt road from Bhurungamari, in Bangladesh's northern Kurigram district in August 2009. Cloud filled, the silver grey sky was reflected in the turgid river's flow. The monsoon – late, erratic, increasingly unpredictable as the climate changes – had finally arrived. The road disappeared beneath a torrent of water. He turned to me and added, 'Rain comes, then river'. The border town – 3 kilometres from West Bengal in India – sheltered from the rain. Jute rope hung over the bridges, jute sticks were stacked in inverted cones. The green jungle shimmered in the humid heat. I was travelling with activist cadres of the Bangladesh Krishok Federation (BKF), a farmers' group and the largest rural-based peasant movement in the country, and the Bangladesh Kishani Sabha (BKS), a women farmers' association. We moved through the rain, mud and jungle. We travelled from village to village, and from meeting to meeting, passing rivers and flooded paddy fields and peasant huts. Rural roads are poorly maintained and bus services are infrequent, making visits by BKF and BKS cadres to villages important organising events. We stayed in the simple homes of the landless peasants and ate fiery fish curry with rice. We drank well water turned muddy red with oxidising iron, and black *chai* (tea) scented with cloves. The meetings sought to develop the organisational power of the BKF, through drawing upon the relational power of village members in order to mobilise landless peasants to occupy land.

Bangladesh is located in the 'tropic of chaos' where the impacts of the catastrophic convergence of climate change, poverty and violence are most acutely felt.[12] It is considered one of the most vulnerable countries in the world to climate change and sea-level rise.[13] Rising sea levels along the coast of Bangladesh are already occurring at a greater than the global average (of 1 to 2 millimetres per year) due to global sea-level rise and local factors such as tectonic setting, sediment load and the subsidence of the Ganges delta.[14] Further, the coastal region is particularly vulnerable to cyclonic storm surge floods due to its location in the path of tropical cyclones, the wide and shallow continental shelf and the funnelling shape of the coast.[15] Some 80 per cent of the country consists of floodplains of the Ganges, Brahmaputra, Meghna and other rivers, which sustain 75 per cent of the country's 160 million people (2011 population).[16] The majority are poor and dependent on agriculture, and are thus more vulnerable to the impacts of changing climatic regimes, particularly flooding.[17]

Since 1987, the national government's policy has been to redistribute *khas* land (fallow or unused state-owned land) to landless households for agricultural purposes, an undertaking known popularly as the Land Law. However, local elites tend to be in control of the land distribution process: local authorities overlook illegal possession of land by large landowners or consolidate their own rights to it.[18]

Since the early 1990s, the government of Bangladesh has implemented structural adjustment programmes, including trade liberalisation regarding agriculture, involving the withdrawal of input subsidies, the privatisation of fertiliser distribution and seed production and the elimination of rural rationing and price subsidies.[19] These have increased farmers' indebtedness and landlessness as they struggle to secure the necessary capital to pay for expensive agricultural inputs.[20] Indeed, functional landlessness (ownership of less than 0.2 hectares) accounts for 69 per cent of the population,[21] brought about through land grabs by rural elites, local government corruption, environmentally induced displacement and government inaction in implementing the Land Law.

For social movements such as the BKF and BKS, the challenges of climate change fold into ongoing conflicts over access to key resources such as land. Land occupation represents an attempt to adapt to the challenges caused by landlessness and climate change, as a BKF activist told me during a meeting in Kathmandu, Nepal, in 2012: 'occupation is our response to climate change, since we cannot rely on the government to help the poor adapt'.

The BKF was established in 1976 and the BKS in 1990. Collectively, they are now estimated to have 1.5 million members. Both social movements are national in scope of operation (their joint office being located in Dhaka) while focused around specific place-based occupations throughout Bangladesh. In efforts to extend their reach, the BKF and BKS also participate in national and international networks of movements. In Bangladesh, these include the Aaht Sangathan (the Eight Organisations, which include migrant labour and indigenous peoples' organisations), and more globally, the international peasant farmers' network La Via Campesina ('The Peasants' Way') and the South Asian Peasant Committee.

Owing to ongoing landlessness and government inaction on implementing the Land Law, the BKF and BKS have, since 1992, organised landless people to occupy approximately 76,000 acres of *khas* land across Bangladesh. The politics and practice of occupation in Bangladesh involves organisational and compositional power: the creation, defence and reconfiguration of material territory into sites of production, social reproduction and potential.

First, through knowing their places of livelihood and habitation, local leaders located in rural communities identify possible sites to occupy and communicate this information to the movement leadership in the BKF/BKS office in Dhaka either through mobile phone conversations or face-to-face meetings held in the capital. For example, occupations have taken place on four islands in the Ganges delta (occupied since 1992), disused railway land (occupied since 2004), swampland water bodies (occupied since 2012) and land inundated by salt water from cyclonic storm surges (since 1998).

Second, the BKF and BKS deploy a cadre of young activists who stay mobile, moving from village to village together with national and local BKF/BKS leaders to generate and reinforce relational power within and between village communities in preparation for land occupation. BKF/ BKS leaders and cadres nurture the organisational power of landless peasants by establishing a coordination committee that mobilises resources (people, skills, finances), the grievances associated with landlessness and the political opportunities provided by the Land Law to shape the 'spatial imaginaries'[22] of peasants: the idea that land is available and they are legally entitled to occupy it. Activist cadres nurture strong relational ties between people (such as trust and interpretive frames) through ongoing place-based face-to-face meetings with communities to enable the mobilisation of resources and people to take the physical risks of engaging in land occupations.

Third, the process of occupation requires an initial politics of intense mobility by peasant farmers through the physical assembly and movement of peasants en masse to secure land by occupying it. In the moment of occupation, armed only with their few belongings, peasant families occupy space by weight of numbers. The 'moment' of land occupation provides the movement with a public presence and helps the development of solidarity between landless peasants.[23]

From an initial mobilisation meeting with a landless community to the physical act of occupation takes from six to twelve months. After the physical act of occupation, the necessities of maintaining it and securing livelihoods necessitate an engagement with the politics of place. The material resources appropriated by peasants are sought after by rural elites, and as a result land occupation faces a counter-movement as peasants are confronted by the potential of violence from and harassment by wealthy landowners and their hired thugs (*goondas*) as well as corrupt district officials. As one BKF cadre told me during a visit to an occupation in northern Bangladesh during 2009: '[P]easant activists are attacked, beaten, burned, jailed and their homes are burned. That is the reality that we face'. As a result the BKF and BKS have developed necessary organising structures, particularly medical and legal knowledge and skills, as another cadre told me:

> For a successful land occupation, the movement needs a strong occupation committee, whose leaders can withstand attacks by the landlords' *goondas*; a strong mass mobilisation; a medical team who can provide medical treatment to those who suffer physical attacks; and a legal team to fight the legal cases brought by landlords in the local courts in an attempt to stop the occupation.[24]

Fourth, to resist physical assault, the occupation must be defended. Peasants told me that they armed themselves with brooms and chilli powder and arranged signal systems to warn the community of impending attacks. The occupiers establish flag signals for communication relays. During the night, an attack is signalled by a hurricane lamp on top of bamboo pole. During the day, a white flag signals a small problem, while a red flag signals a *goonda* attack. The BKF and BKS have also stayed mobile by organising the simultaneous occupation of five islands in order to spread landlord and *goonda* resources thinly across multiple locations. Successful defence of an occupation enables peasants to commence the process of reconfiguring space and power relations – that is, making space.

Land occupation constitutes places as sites of social reproduction wherein the gendered agency of women is paramount: the logistics of where to sleep, eat, wash and defecate precede the construction of homes, the growing of crops (such as rice, vegetables, pulses and fruits) and the sheltering of animals (primarily cows). Through the process of occupation, peasants attempt to construct sites of potential: a movement against the interests of agrarian elites and the state (and market-led agrarian policies) in favour of the production and reproduction of family labour and towards more equitable and ecologically sustainable practices.

For example, the BKF/BKS argue for the importance of food sovereignty practices, as noted by the president of the BKF, Badrul Alam, at a South Asian regional conference of La Via Campesina held in Dhaka in 2008:

Food sovereignty means the people's right to produce and consume culturally appropriate and accepted healthy and adequate food and their right to define their own food and agriculture policy ... It prioritises the local and national economy, peasants and family farm-based agriculture, artisan-style fishing, pastoralist-led grazing and food production, distribution and consumption based on environmental, social and economic sustainability.[25]

In such agricultural sites of production, the reconfiguration of power relations (away from the state and capital towards poor peasants) that food sovereignty promises is potentially an important force of momentum for social movements occupying land. While definitions of food sovereignty vary between organisations and activist networks, have changed over time and contain inconsistencies, common themes have emerged such as direct peasant participation in agrarian reform, including peasant control over territory, biodiversity (commons) and the means of (food) production. Food sovereignty farming practices are employed in an attempt to repair the dynamic and interdependent process that links society to nature though labour that has been undermined by the exploitation of society and nature through capitalist agriculture.[26] They also enable peasant communities to mitigate and adapt to the effects of climate change because of the biological resistance of crops, the recovery capacity of land and the interdependent social dynamics among peasants.[27]

As sites of potential, attempts to practice food sovereignty are precarious. In each of the occupations I have visited, everyday life has been an ongoing struggle against dispossession by landlords, the vagaries of the weather (especially as the climate changes) and attempts to fashion sustainable livelihoods under conditions of relative poverty. As one BKS activist told me on the occupied island of Charhadi:

> On some islands, people have been dispossessed of their land by landlords from the mainland ... [D]espite successfully remaining on the island for ten years, people still have no education or health care, and no flood shelters for their cattle when the river floods during the monsoon. Since the occupation, nearly one hundred, mostly children, have died.[28]

Nevertheless, despite these hardships and difficulties, the BKF and BKS continue to represent the best means for landless peasants to transform their lives. Over the last 25 years, land occupations organised by the BKF and BKS have enabled the distribution of land to more than 107,000 of the poorest men and women living in the countryside. In addition, the BKF and BKS demand that the Bangladeshi government adopt food sovereignty as a national policy, hoping to galvanise representational power to achieve their goals. In 2011, the BKF president was invited by the government to participate with the Bangladeshi delegation at the meeting of United Nations Framework Convention on Climate Change in Durban, South Africa, thus opening up a potential site of collaboration. Further, the BKF and BKS have been able to extend their reach through organising 'climate caravans', which I will discuss in Chapter 6. While Bangladeshi peasants occupy space to fashion livelihoods, space may also be occupied in order to articulate systemic protest, as with the Occupy movement.

OCCUPY AND URBAN COMMONS

During 2011, a wave of mobilisations and occupations of public squares took place around the world (particularly in the United States, the UK, Spain and Greece), protesting against the austerity politics ushered in by governments after the financial crisis of 2008, as well as long term unemployment. The protests were inspired by the mobilisations of the Arab Spring of 2010/11, such as the occupation of Tahrir Square in Cairo, Egypt.

The Occupation of Tahrir Square

Inspired by the revolution in nearby Tunisia, what became known as the Arab Spring spread to Egypt in 2011. In a country with increasing levels of poverty, a 25 per cent youth unemployment rate, low wages, underemployment and widespread government corruption, protests and strikes began to emerge in opposition to the military regime of President Hosni Mubarak. Protestors represented a diverse class base that included the young unemployed (of whom there were 1 million aged between 20 and 24 years), students, miners, bankers, transport workers, factory workers and teachers.[29] In particular, they struggled against the political and economic exclusion that was a result of neoliberal structural adjustment programmes and the state violence that accompanied their enforcement. These programmes had resulted in the privatisation of public institutions, the scaling back of social safety nets, the curtailment of workers' rights and the reduction of living standards.[30]

Initially, groups of activists stayed mobile. Circulating around the city for three days, fighting with police and burning down police stations, activists gathered in the protected spaces of alleys before emerging onto Cairo's streets to join together in mass street actions culminating in the mass swarming and occupation of Tahrir Square. Protestors knew their place in choosing Tahrir Square as a site of resistance to the Mubarak regime. Tahrir (or Liberation) Square was so named after the 1952 Egyptian revolution and was a potent symbol of Egyptian nationhood and rebellion, having been the focal place for many protests and demonstrations since the earlier 1919 revolution. It is also a site of decision-making, being the location of the Mogamma government administrative building and the headquarters of the Arab League.[31]

Representing a key site of circulation, Tahrir Square, once occupied, became the physical and symbolic hub of the Egyptian uprising. Activists deployed compositional power and made space by establishing a site of social reproduction including a tented encampment, with electricity rigged from street lights, street hospitals, waste and recycling stations, a prison, supervised day-care facilities, food stalls, barricades, stages, and a microphone and loudspeaker enabling speeches, news and debates to take place during open mic sessions that waged wars of words against the regime.[32] Approximately 847 people died and 6,000 were injured during the 18 days preceding the resignation of Mubarak. Subsequently, democratic elections were held in which the Muslim Brotherhood's Mohamed Morsi came to power as president in June 2012. However, this government was short-lived. One year later, a military coup saw the constitution suspended, a crackdown on the Muslim Brotherhood and arrests of those who protested against the coup. However, despite such setbacks, Egyptian revolutionaries have succeeded in disrupting the relationship between the Egyptian state and its citizenry. Ongoing protests see farmers resisting the privatisation of their land, workers seizing control of sites of production such as factories, and Bedouins occupying a government site of destruction, a nuclear site, to reclaim appropriated territory.[33]

Occupy

The catalyst for the Occupy movement was the 'squares' protests initiated by the Indignados, a protest movement that emerged in Spain in May 2011, mobilised by the digital platform ¡Democracia Real Ya! (Real Democracy Now!). The protests spread to 50 Spanish cities, and subsequently to Greece (where protestors set up an encampment and occupied Syntagma Square in front of the parliament building in Athens) and to Turkey (where protestors occupied Gezi Park in Istanbul). These protests inspired US activists to occupy Zuccotti Park, located in New York City's financial district (near the New York Stock Exchange), in what became known as Occupy Wall Street. The initial call to occupy Wall Street appeared in the July 2011 issue of the Canadian magazine Adbusters, which cited Tahrir Square in Cairo as its inspiration. Wall Street symbolised the influence of financial corporations on US political, economic and social life.

Occupy necessitated the physical taking of space, and practised what Jeff Juris has termed a 'logic of aggregation': the mass assembling of folk in public places.[34] The occupation of space by protestors gives that space a physical presence, a locational identity that can be associated with a protest. It also enacts spatial symbolism, since locations carry meanings, and the choice of location symbolises something about the protest.[35]

The protest tactic of Occupy acted as a meme – a unit of culture (such as buzz words, fashion trends, symbols, ideas) that spreads beyond its creator. Occupy extended its reach facilitated by social media (Twitter, Facebook, blogs and so forth) and smartphones that allowed individuals to continually post and receive updates and circulate texts, videos and images. In addition, existing activist networks and interpersonal ties further facilitated the spread of the protest meme. This information spread because the grievances in one place – frustration with 'politics as usual' and a lack of trust in the usual political actors – resonated in other places as well.[36] A wave of occupations took place across US cities and generated 1,500 protests in 82 countries, including events in London, Amsterdam, Brussels, Buenos Aires, Cape Town, Hong Kong, Lima, Oakland, Oslo, Madrid, Manila, New York, Rome, São Paulo, Seoul, Stockholm, Taipei, Tokyo, Toronto and Zurich.[37]

Occupy took place at sites of decision and circulation. For example, Wall Street and the City of London are key sites of corporate financial decision-making and the circulation of capital, while the state legislature in Wisconsin is a key site of state government decision-making. In Occupy Oakland, protestors worked with the Longshore and Warehouse Union to close the port of Oakland.[38] Occupy also challenged the meaning of spaces by symbolising their exclusionary character: how parks, squares and the streets are rarely 'public' but are governed by laws and frequently encode inequalities. For example, the City of London was a focus of excessive capital accumulation at a time when the majority of the UK population were being subjected to austerity policies.[39]

Drawing on compositional power, protestors made space through the construction of tent encampments with kitchens, bathrooms, libraries, first-aid tents, information centres, sleeping areas and educational spaces (such as 'Tent City University' in London). As sites of social reproduction, these spaces also saw activists establishing resources for themselves rather than drawing upon traditional (state) services. For example, in addition to providing food for the occupiers (via kitchens and donated food), mental health care, drug counselling (through medical posts) and temporary housing was provided.[40]

In this way the reproduction of everyday life was also a form of political activism, what Silvia Federici terms a 'reproductive commons':

> if 'commoning' has any meaning, it must be the production of ourselves as a common subject. This is how we must understand the slogan 'no commons without community' … community as a quality of relations, a principle of cooperation and responsibility to each other.[41]

Occupy contrasted the public space owned and controlled by the state with a common space, opened and shared by those who occupied and shared it according to their own rules.[42] In addition, the relational power between protestors was developed. 'Being-in-common' generates important relations of solidarity between protestors as a condition for new political possibilities.[43] For example, in the large popular assemblies at Occupy Wall Street, people communicated via the 'people's microphone' in order to circumvent a ban on the use of bullhorns without a police permit. Someone addressing a mass meeting would pause after each phrase and the people nearby would repeat it in unison to the crowd. Given that anyone (in theory) could speak, it was potentially empowering for people to hear their views repeated and amplified by others.[44]

As sites of potential, Occupy protests not only represented a rejection of representative politics, they also embodied prefigurative politics. This is because, in the spaces of Occupy, novel forms of collective self-organisation, consensus decision-making and popular assemblies were practised as attempts at enacting direct democracy. Such forms of 'horizontality' are comprised of dynamic social relations based upon an affective and trust-based politics that attempts to be non-hierarchical.[45]

The Occupy protests also embodied a war of words. First, the very word 'occupy' acknowledges that the act of occupying space symbolises the need to reclaim it from corporate greed and what was deemed unrepresentative politics. Second, protestors were emotionally driven by anger at the immense economic inequalities wrought by finance capitalism and adopted the slogan 'we are the 99%'. This was deployed to act as an inclusive and majoritarian notion of 'the people', posited against the excessive wealth of the 1 per cent minority, thereby highlighting the inequalities inherent in the capitalist system.[46] In challenging common-sense assumptions about the need for a political economy of austerity, and the normalisation of economic inequities, Occupy protests also represented sites of assumption.

However, despite the potential inherent in the practice of prefigurative politics, the Occupy protests were beset with a range of problems. Certainly in Europe and the United States, the majority of protestors were white, educated and middle class, raising important issues about the exclusionary character of the protests, which were unable to represent those already excluded by neoliberal economics. This was compounded by the construction of a unitary political subject that effaces difference characterised by the slogan of 'the 99%'. There was no strong anti-oppression platform that addressed racism, sexism, homophobia and so on.[47] For example, there were frequent reports of gender discrimination (for example, in the male-dominated general assemblies and in the division of labour at the encampments), and there was sexual harassment of women at many of the Occupy camps.

In addition, working people were unable to live at the camps full time or often attend assemblies (given that assembly timings were often during people's normal working hours). This was compounded by a failure – except in Occupy Oakland – to fully engage labour unions in the Occupy struggles. The focus on occupying space in protest against financial and state exclusion also gave little consideration to indigenous people, who had historically been spatially dispossessed.[48] Finally, Occupy's decision-making body, its general or popular assembly, was unable to fully address the everyday needs of the encampments, and created bottlenecks in leadership that actually prevented organisers from making decisions that could have kept the movement alive.[49]

Many of the internal problems of Occupy (internal conflicts, hierarchies, burnout) arose out of the logic of aggregation – a focus on assembling large numbers of protestors in one place and an ongoing commitment to maintain the intensity of the occupation over an extended and indefinite period of time.[50] Against police violence and the harshness of winter weather in many places, this proved unsustainable. Indeed, it has been argued that a fetishisation of space occurred whereby the occupation of space became an end in itself at the expense of broader issues concerning economic inequality, social injustice and the power of the corporate sector.[51]

In addition, Occupy privileged prefigurative politics over strategic politics. A focus on strategic politics would have constructed a mandate within the occupations (and between them) for strategic political intervention (and perhaps structural change). However, a focus on prefigurative politics tended to view decision-making processes and the physical occupation (and daily life) of the encampments as ends in themselves, rather than tactics within a larger strategy of political contestation. In addition, self-organisation within the camps served to legitimise the withdrawal of state provision of key resources and services that has attended neoliberal privatisation. In short, Occupy became more about the act of occupying than of taking on the dominant powers of the finance and state sectors.[52]

However it revitalised the wave of protests associated with alter-globalisation during the late 1990s and 2000s (see Chapter 6). Occupy moved from the plazas to continue in neighbourhood assemblies and workplaces, and generated a range of new struggles including organising against evictions in Spain and against debt in New York City.[53] In Greece, such a move has generated experiments in urban commons.

Urban 'Commons' in Greece

Following the global economic crisis of 2008, Greece experienced a 25 per cent drop in gross domestic product between 2008 and 2013. This was accompanied by the closure of over 100,000 businesses from 2011 onwards, an unemployment rate of 27 per cent by 2013, and the number of people living below the poverty level reaching 44 per cent.[54] This has been exacerbated by ongoing brutal austerity measures imposed on the country by the so-called Troika of the International Monetary Fund, European Central Bank and European Commission, which has led to the privatisation of public services and public spaces and the undermining of state pensions.

Following protest encampments in Syntagma Square in Athens and elsewhere in Greece in response to this situation, there has been a flourishing of Occupy-style practices in urban neighbourhoods. Through necessity, residents of neighbourhoods have had to focus on everyday practices of activism that are examples of urban commons, and that act as sites of production, social reproduction and potential. For example, in Thessaloniki, workers have restarted the Vio.Me factory, shifting production from tile-jointing compounds to disinfectant gels that have been supplied to dispensaries within the broader anti-austerity movement.[55]

As Athina Arampatzi argues, in Athens over 300 neighbourhood groups have developed that draw upon the compositional power of urban residents as well as deepening the relational power between them. For example, in the Exarcheia neighbourhood during 2013 and 2014, community groups, residents and individual activists drew upon their knowledge of place and attempted to address neighbourhood decay and unemployment through resident committees that have reclaimed public spaces and attempted to re-signify their use.[56]

Arampatzi argues that three main initiatives emerged from this. First, in waging wars of words, alternative narratives have been produced that reframe local problems as outcomes of the financial crisis and the role of the neighbourhood within broader (anti-austerity) struggles. Second, residents have deepened their relational power, developing new cooperative interactions between themselves as they participate in a range of neighbourhood initiatives that act as sites of social reproduction, prioritising social needs over profit-making. These have included initiatives that attempt to practice different forms of 'solidarity economics' including 'without middlemen' markets, non-monetary and alternative currency exchange networks, time banks, community cooking collectives, social centres, work cooperatives and health clinics. Third, through the deployment of alternative forms of organisational power, experiments in more participatory and autonomous forms of decision-making such as open assemblies, neighbourhood assemblies, collective organisation and factory recuperations act as sites of potential. Importantly, people have also attempted to extend the reach of these practices beyond Exarcheia, linking up with communities elsewhere in Athens and with broader struggles against resource exploitation throughout Greece. For example, the Assembly for the Circulation of Struggles includes activists from a range of campaigns that exchange experiences and co-produce written materials for movements across the country.[57]

MAKE SOME SPACE

'Making some space' is first and foremost about the transformation of the character of, or meanings associated with, places. It also concerns the productive relations and practices that emerge in cultures of resistance, which frequently contain prefigurative political processes. However, as we have seen in the examples discussed above, there are interesting differences in the role of this kind of politics.

In Pollok Free State, the strategic goal of the protest was structural: to stop the construction of the M77 motorway extension. Prefigurative practices formed part of the everyday life of protest rather than its raison d'être. In contrast, in the Occupy camps, protests lacked a primary structural goal of social change, and their occupation of space was fetishised rather than used as a strategic tool of the struggle. In Athens, the everyday prefigurative politics is in part structural: the necessity of fashioning social reproductive alternatives as a response to the deprivations of austerity policies in Greece. In Bangladesh, land occupation and prefigurative practices such as food sovereignty form an integral whole of political strategy and practice aimed at structural change by challenging and attempting to remedy poor landless peasant exclusions in Bangladeshi society. In all cases however, making space is a response to perceived injustices and inequities that frequently also require forms of strategic mobility, the issue to which the next chapter is devoted.

4

Stay Mobile

Packs and Swarms, Flash Mobs and Hacktivism

'Stay mobile' refers to how strategies of mobility – movement in and across space – are critical for the prosecution of protest. Such strategies of mobility generate their own, different geographies. 'Stay mobile' also refers to the necessity of activists to continually adapt to changing contexts and conditions.

Protestors manifest varied and interrelated spatial mobilities. As I discussed briefly in Chapter 3, practices of making space are interwoven with strategies of mobility in many forms of protest. Such strategies of mobility can generate different relationships to space: space may be claimed, defended, strategically used and/or abandoned depending upon the strategies and goals of a particular protest. These strategies are potentially fluid and diverse, continually adapting to different physical terrains and to changing circumstances.

The spatial practices associated with staying mobile necessitate an interplay between movements involving the territorialisation and deterritorialisation of space exemplified by the pack and the swarm.[1] Packs are small in number, often constituting 'crowd crystals' that may precipitate crowds, demonstrations and social movements.[2] The pack does not openly confront dominating power; it is more secretive, utilising underground or guerrilla tactics, surprise and the unpredictability of constant movement. Packs effect a deterritorialisation of space – they tend to move across space rather than occupying it. Their action always implies an imminent dispersion.

The swarm, by contrast, is large in number, effecting a movement of territorialisation. The swarm openly confronts dominating power by weight of numbers, by occupying space – be it physical, symbolic, political or cultural. Swarmings can be either predictable or not, sometimes a prelude to, or an expression of, social movements. At other times swarms may articulate spontaneous moments of defiance, anger or elation that decompose almost as soon as they form.

The spatial practices of the pack and the swarm may be interrelated or opposed, a prelude to, or transformation of, the other. Their relationships to space – those of territorialisation and deterritorialisation – are always relative, always interwoven with one another.[3] For example, mass street actions such as demonstrations take over – or territorialise – particular spaces in the city. They are frequently composed of numerous packs of activists representing particular organisations, groups and individuals, who come together to form a demonstration. The protesting swarm is mobile, snaking its way through city streets, often targeting sites of circulation and thereby disrupting business as usual (such as traffic and commuter flows); they frequently come to rest (such as in a rally) and take control of urban space before dispersing. At root in such processes, and enhancing strategic mobility, is the politics of affinity.

THE POLITICS OF AFFINITY

Protests bring together a variety of organisations, groups and individuals in a broader affinity that frequently traverses gender, age and class differences. Within this broader affinity are frequently smaller groups called affinity groups. Affinity groups take their name from the *grupos de affinidad* devised by Spanish anarchists during the 1930s. They were subsequently popularised during the 1970s in the non-violent resistance against a nuclear reactor in Seabrook, New Hampshire. During the 1980s, the peace movement in the UK and United States utilised affinity groups for both specific actions and for ongoing, small and supportive action groups. Affinity groups have subsequently formed an important component of the anarchist and autonomist-inspired direct action involved in anti-roads protests (see Chapter 3), alter-globalisation protests (see Chapter 6) and climate justice activism (see below).

Numbering between 3 and 15 people, affinity groups consist of a group of people who share common ground (friends, lovers, shared sexualities, ethnicities, religious beliefs and so forth) and can provide supportive, sympathetic spaces for their members to listen to one another, share concerns, emotions, fears etc. In affinity groups, people provide support and solidarity for one another. Small enough to enact a consensus form of decision-making, these groups are non-hierarchical and participatory, embodying flexible, fluid modes of action. Affinity groups will frequently join together with other affinity groups for political actions, but retain their own separate integrity and course of action within the context of broader struggles. The common values and beliefs shared by these groups constitute what Raymond Williams has termed a 'structure of feeling' resting upon collective experiences and interpretations.[4] Within affinity groups and between them, protestors seek to act upon common ground.

Indeed, the politics of affinity – if not affinity groups – underpins all forms of protest, irrespective of the structures of political organisation. The importance of affinity for building solidarity in mass-based movements is critical, even where more hierarchical organisational logics predominate.

To discuss these processes I will first consider the Maoist Naxalite movement in India. The Naxalites provide an example of the interplay between the swarm and the pack and their different spatial mobilities under conditions of armed struggle. Because the Naxalites operated during the late 1960s and early 1970s (with the associated use of pre-digital technologies), I then consider some contemporary examples. Hence, I examine the Black Lives Matter mobilisations in the United States. Unlike the political party-based Naxalites, Black Lives Matter is a network-based mobilisation, grounded in communities protesting against police violence on people of colour, and enabled by the use of digital media. Next, I examine the flash mobs of Idle No More, an indigenous movement that combines traditional cultural practices with social media tools. Finally, I consider forms of hacktivism that utilises the swarm intelligence of social media to effect forms of culture jamming by activist packs.

THE PACK AND THE SWARM IN THE NAXALITE MOVEMENT[5]

Hark!
Listen to the call of the rebels.
Come out and join them.
March forward and break the chains of servitude.[6]

The Naxalite movement took its name from the village of Naxalbari in West Bengal where, in 1967, Maoist dissidents within the Communist Party of India (Marxist), also known as the CPI(M), first organised an insurgent struggle against landlords. Peasants, sharecroppers and landless wage labourers had been systematically impoverished by the Indian agricultural system. Poor peasants were forced to mortgage and later sell their land to big landlords, thereby becoming tenants and sharecroppers.[7]

Many tenants were exploited by *begar* (forced work for their landlord) and the imposition of levies to make them bear the costs of ceremonies in their employer's house on special occasions. Landless agricultural labourers – who numbered 45.4 million in India in 1971 – suffered not only from low wages but also from underemployment, and because of this were forced to borrow money from private moneylenders at exorbitant rates of interest.[8] By 1968, it was estimated that 54 per cent of India's 230 million rural population existed below the minimum standards of living as defined by the Indian government.[9]

Communists and Santhal *adivasis* (tribals) mobilised organisational power and initiated a movement in sites of production to forcibly occupy agricultural land and expropriate rice hoarded by landlords. Naxalbari provided the opportunity for Charu Mazumdar, a CPI(M) activist working in the area, to put Maoist theories of armed struggle into practice.[10] Although the Naxalbari uprising lasted only a few months – it was suppressed by the West Bengal government's police forces – its consequences were dramatic as the Naxalites extended their reach.

First, across India, there was a dramatic increase in land occupations by peasants, demonstrations demanding land for the landless, agitation for increases in the wages of agricultural workers, the forcible harvesting of crops by sharecroppers and protests against taxation. These were organised by communist dissidents who were inspired by the events of Naxalbari and calls for waging revolutionary armed struggle articulated by the Naxalites.

Second, as well as economic demands, struggles led by communist dissidents for the seizure of political power also commenced. With the events in Naxalbari acting as a catalyst, these struggles tended to emerge in those areas where peasants and tribals had been politically organised by communist cadres for many years previous to the Naxalbari rebellion. Although organised around place-specific economic, political and cultural issues, these struggles allied themselves to the revolutionary communist cause, adopting most of the Naxalites' ideology, political organisation and strategy (particularly that of armed struggle). These revolutionary groups also became affiliated to the Communist Party of India (Marxist-Leninist), or the CPI(M-L) – the Naxalites' political party – when it was created in 1969.

Between 1968 and 1971, guerrilla packs emerged in various districts in the states of Andhra Pradesh and West Bengal with the aim of capturing political power in those states. Guerrilla activities took place in both the towns and the rural villages surrounding them. Revolutionary committees were also established, setting wage rates, redistributing land, cancelling debts and organising village defence groups. Elsewhere, Naxalite activities emerged in other Indian states, including Assam, Nagaland, Orissa, Uttar Pradesh, Himachal Pradesh, Tripura and Punjab. Many of these activities were temporary in nature, ranging from broadcasting Maoist propaganda through posters and pamphlets to isolated killings of landlords and the snatching of guns.[11]

Naxalite strategy was based upon three aspects of organisation. First, it was conspiratorial and underground, although not all packs went underground at the same time. Second, it avoided the development of open mass organisations. Third, there was a separation of guerrilla warfare activities from the regular party machine, the party having no control over recruitment to the guerrilla packs.[12] The movement uncritically adopted Maoist ideology but departed from Maoist political strategy in several ways. First, as with Maoist strategy, the Naxalites celebrated the spontaneity of the peasantry, but placed this above all other tactical considerations. Second, as with Maoist strategy, the Naxalites considered the peasantry to be a revolutionary force, but did not act as if it were such a force – control of Naxalite activities both in the central leadership and at state and local level was assumed by middle-class cadres and party intellectuals. Finally, the Naxalites were not able to establish a united revolutionary force (as Mao had done in China), and revolutionary sites were isolated and often not in communication with the party or each other – that is, they failed to fully extend their reach.

Naxalite strategy placed great emphasis upon the spatial mobility of guerrilla packs (*dalams*) within the specific locales of action (such as a particular village). The utilisation of packs by the Naxalites emerged through the evolution of the tactic of the annihilation of class enemies (*khatam*), which became equated with the uprooting of the entire feudal class structure. The *khatam* line operated through an underground structure, exemplified by the slogan 'one man, one village, one action', and would involve a Naxalite cadre infiltrating a village, setting up a covert pack of four to five peasants, and proceeding to kill a landlord.

The assassination of a landlord or moneylender was expected to act as a catalyst whereby the victim's crops would be seized, the peasants would lose their fear of their oppressors (having seen how they could be destroyed) and, realising their relational and organisational power, the peasants would be inspired to join the struggle. Hence an individual action by a pack was expected to lead to mass action.[13] The strategy of the pack interwove political ideology with a specific spatial practice, that of numerous, autonomous packs, operating in secret, constantly on the move, carrying out localised assassinations in an unpredictable manner across a territory. According to one estimate, at the height of the movement, up to 700 guerrilla packs were operating in West Bengal alone.[14]

In some of the main sites of struggle, Naxalites also adopted the spatial strategy of the swarm, the territorialisation of physical, symbolic, political and cultural space by mass of numbers as an open confrontation with the state. The various localised insurgencies proceeded in four ways. First, by the physical territorialisation of sites of production (such as agricultural land), including the forcible cutting of crops, thefts of grain and other property from landlords in the area and the seizing of documents detailing loans given to tribals by landlords. Second, by the symbolic territorialisation of space through establishing peasant 'ownership' by ploughing small parcels of land. Third, by the political territorialisation of particular villages, making space through establishing parallel administrations (as a form of representational power) such as setting up village peasant committees, redistributing land amongst landless peasants, cancelling debts owed by peasants to landlords and setting up people's courts to dispense revolutionary justice to landlords and moneylenders.[15] Fourth, by the attempted cultural colonisation of tribal and peasant culture by Naxalite ideology whereby primacy was given to agrarian class relations at the expense of ethnicity and ecology (that is, there was an absence of Naxalites knowing their place). The initial success of these swarmings was exemplified by the experience of the Srikakulam insurgency in Andhra Pradesh, where the movement claimed that red political power had been established in 300 villages before the movement waned.[16]

The processes of territorialisation and deterritorialisation in the Naxalite movement were always interwoven. For example, even though constantly moving, packs would temporarily establish themselves in a village long enough to enact the *khatam* or, after conducting an assassination, pack members might participate in land occupations. Likewise, if confronted by a superior military force, a land or village occupation or swarm might disperse rather than remain to fight. Ideologically, the actions of a pack were often seen as a prelude to that of a swarm, based on the idea that an assassination would inspire mass revolt. Meanwhile, swarmings, if dispersed or defeated, might coalesce elsewhere as several spontaneous packs.

In response to the Naxalite movement, the central and state governments deployed armed police, paramilitary forces and the Indian army in a concerted programme of counter-insurgency – an 'encirclement and suppression' policy, which effected a reterritorialisation of perceived 'Naxalite space'. Entire villages were surrounded by police, huts were raided, people arrested and villages burnt to the ground. Any communist leader who was captured was shot dead. In response to this, the movement stepped up its assassination campaign and stressed underground activities by packs over open mass actions by swarms.

In conjunction with this process, the central government introduced a series of draconian laws severely curtailing civil liberties, exemplified by the Maintenance of Internal Security Act (MISA), introduced in May 1971, under which people could be detained without trial for an indefinite period.[17] Once imprisoned, activists were liable to be murdered.

As various localised movements experienced ruthless repression, the rural bases were eroded and a new area of struggle emerged in the streets of Calcutta, the state capital of West Bengal, during 1970/71. However, this too was brutally repressed by the Calcutta police and CPI(M) forces.[18]

Subsequently, Naxalite resistance re-emerged in Bhojpur district in Bihar and Telengana district in Andhra Pradesh in the mid 1970s. Maoist rebellion, under the Communist Party of India (Maoist), has continued to the present, and has subsequently spread to various parts of eastern India in what is known as the 'Red Corridor'.[19] The Naxalites remain a significant form of peasant and indigenous resistance to government and corporate attempts to force the poor from their land (frequently through the use of violence) in order to exploit minerals and construct industrial developments.[20] While the Naxalite movement was a movement organised by the Communist Party and effected by packs of activist cadre, Black Lives Matter represents a more network-based form of social protest using media unavailable to the Naxalites.

BLACK LIVES MATTER AND COP WATCH

African American and Latino people in the United States are subjected to a regime of state violence, termed the 'racial state'.[21] Racialised difference is both produced and reified through a range of processes that are in part mediated by state institutions that result in multiple systemic oppressions being visited upon people of colour. Black dispossession and internal colonialism have been materialised through the spatial segregation of the 'black ghetto'.[22] This was constructed through deliberate policies of urban renewal by the state, banks and real estate firms that contained black people in small, overcrowded areas with poor housing stock in inner city areas. Subsequently, as the inner city was gentrified, black residents got displaced to black suburbs such as Ferguson, Missouri (which was 69 per cent black in 2010). Both inner city and suburban areas have been subjected to aggressive policing and other forms of state violence.[23]

This violence has included the killing of black people by law enforcement officers, police brutality, racial profiling and surveillance, racial inequality in the US criminal justice system (where 60 per cent of all those incarcerated are people of colour), the employment of stop-and-search policies on young black males, unemployment and the withdrawal of state resources for welfare support. In what has been termed the 'prison industrial complex', the mass incarceration of people of colour has seen a proliferation in the construction of prisons (and the associated profits generated by this) as well as the generation of a cheap workforce for a range of corporations through what is termed 'insourcing'.[24] For example, corporations such as Wholefoods, Walmart and McDonalds have employed incarcerated people who earn on average between $0.23 and $1.15 per hour, six times less than the federal minimum wage in the United States.[25]

The popular response to this oppression has taken the form of Black Lives Matter (BLM), a chapter-based national activist network that emerged in the United States in 2012 within the African American community. It began when three Black queer women – Patrisse Cullors-Brignac and Alicia Garza of the National Domestic Workers Alliance, and Opal Tometi of the Black Alliance for Just Immigration – initiated an online social media campaign with the use of the hashtag BlackLivesMatter in response to the acquittal of George Zimmerman in the fatal shooting in Florida of Trayvon Martin, an African American teenager. The idea of the online platform was to connect people through dialogue, collective action and protest concerning racism.[26] BLM has subsequently protested against various forms of state violence that African Americans and Latinos are subjected to.

BLM has extended its reach through connecting up black struggles in the US with other struggles in the country as well as abroad. BLM works inter-sectionally, incorporating the struggles of black women, men, LGBTQ people, disabled people and undocumented black immigrants. In so doing, BLM draws connections between issues of racism, economic inequality, immigration law, homophobia and disability within contemporary US society:

> When we say Black Lives Matter we are talking about the way in which Black people are deprived of our basic human rights and dignity ... Black women continue to bear the burden of a relentless assault on our children and our families and that assault is an act of state violence. Black Queer and Trans folk bear a unique burden in a hetero-patriarchal society that disposes of us like garbage ... [T]he fact that 500,000 Black people in the U.S. are undocumented immigrants and relegated to the shadows is state violence ... Black folks living with disabilities and different abilities bear the burden of state-sponsored Darwinian experiments.[27]

BLM acts as a site of assumption, challenging the status quo of contemporary US society that has normalised state-sanctioned violence in its various forms, such as fatal shootings of black people by police and the armed-force doctrine of the police. This has seen the militarisation of police tactics and equipment since 2006, with $4 billion in military hardware flowing from the Pentagon to the police, and the use of live ammunition, tear gas, smoke grenades, armoured vehicles and curfews.[28] In opposition to this, BLM has frequently deployed organisational power and adopted the tactics of the pack and the swarm.

A dramatic example of this occurred during the protests that followed the fatal police shooting of Michael Brown, a black 18 year old, in Ferguson, Missouri, on 9 August 2014. BLM organised a Freedom Ride, mobilising packs of activists from across the United States (from such places as New York City, Boston, Chicago, Detroit, Los Angeles and San Francisco) to converge in a protest swarm on Ferguson. Packs of activists made some space by blocking the entrances to a site of decision, conducting die-ins in front of the federal courthouse. Activists also intervened in sites of circulation. For example, a mass swarming action shut down 'the built landscape of circulation' by blockading junctions to the interstate highway.[29] Two weeks of protests including acts of arson and rioting were followed by two further waves of protest. The second took place in Ferguson in November 2014 when the police officer responsible for Brown's shooting was acquitted, while a third took place in 2015 on the anniversary of the shooting.[30] On all three occasions, activists knew their place. Ferguson was not only the site of a fatal shooting of a black teenager, it was also the site where city officials had directed the police department to raise revenues for the city budget through 'broken windows' policing tactics that targeted small-scale offences with fines that disproportionately targeted poor people of colour.[31]

BLM has also waged a war of words, demanding community based alternatives to policing and prison, the reduction of school surveillance and an end to police brutality. Their slogans – 'Black Lives Matter', 'I Can't Breathe' and 'This Stops Here' – speak to the oppression and violence faced by people of colour and the discrimination they experience at the hands of a racial state as well as their resistance to it. More recently, through use of social media such as Twitter and the Black Lives Matter website, BLM mobilisations have extended their reach, taking place simultaneously across six cities (Washington, Los Angeles, San Francisco, Chicago, Chattanooga and Minneapolis) to protest against police shootings of black people and discriminatory police profiling of black people at airports, shopping malls and on the streets of cities.

In Minneapolis, a BLM swarm protested in a site of consumption, the Mall of America – drawing attention to the privatisation of public space – as a diversion to another planned event at a site of circulation, the Minneapolis–St Paul airport. While the demonstration at the mall took place, other BLM packs were organising blockades of the roads leading to the airport terminals. After a brief demonstration, the swarming at the mall quickly dispersed, as activists boarded trains to the airport to join the road blockade.[32] BLM has been able to stay mobile, keeping strategically ahead of the police force, and has used multi-site communication to generate momentum and mobilise across the United States.

Subsequently, BLM has extended its reach beyond the United States as it has begun to forge connections with Brazil's informal communities and the Palestinian struggle.[33] Further, BLM has emerged in the UK, where in August 2016, BLM events took place in London, Birmingham and Nottingham to protest against the racial state in the UK. Activists intervened in sites of circulation, blocking streets and city centres to bring traffic to a halt, and blockading roads to Heathrow airport. The latter was done because BLM activists in the UK also knew their place.

Heathrow airport had been the site of the death of Angolan deportee Jimmy Mubenga, who died while being restrained by private security guards in 2010.[34]

BLM activists in the United States have also been involved in Cop Watch – a dispersed network of activists who patrol their neighbourhoods in packs using cameras to monitor and document police behaviour. The presence of Cop Watch packs on the streets has changed police behaviour, ameliorating their use of violence against black people. The key to Cop Watch's success has been their knowledge of place, rooted in neighbourhoods, where they have developed relational power and obtained community support and information on police movements. Their organisational power has enabled them to stay mobile by mixing their repertoire, for example by walking different routes to keep the police guessing where they are located.[35] While BLM is a network-based mobilisation, in the following example of Idle No More I show how contemporary social media interacts with cultural tradition in the fashioning of protests.

FLASH MOBS

Another form of staying mobile is the flash mob. A flash mob is an unrehearsed, spontaneous, contagious and dispersed mass action, such as an impromptu demonstration. Groups of people congregate in public spaces to engage in particular acts before dispersing again. Flash mobs are organised through virtual and viral means (such as e-mails, text messages, word of mouth, blog posts, social media platforms) and display a capacity for self-organisation and 'swarm intelligence'. They are a form of dispersed collective knowing that is based on sharing and communication without the need for centralised control.[36]

For example, UK Uncut have used affinity groups and flash mobs in response to the UK government's massive cuts to public spending as part of its austerity policies following the financial crisis of 2008. In 2010, 70 UK Uncut activists – representing several affinity groups – swarmed a site of consumption, the Vodaphone store in London, and occupied it to draw attention to accusations that the company was involved in tax avoidance. The idea extended its reach, so that within three days over 30 Vodaphone stores had been shut down around the country by flash mobs organising over Twitter.[37] Subsequently, since 2010, 800 actions have taken place across the UK, wherein activists have made some space by occupying sites of production (of profit) such as banks and transforming them into sites of social reproduction such as crèches, classrooms, food banks, homeless shelters and libraries.[38]

A further example of the use of flash mobs comes from Idle No More (IDM), a Canadian grassroots movement for indigenous rights, sovereignty and environmental justice. IDM emerged in late 2012, launched in Saskatchewan by four women, Nina Wilson, Sheelah Mclean, Sylvia McAdam and Jessica Gordon, at a teach-in to discuss a series of legislative changes (Omnibus Budget Bill C-45) that would remove environmental and legal protections for forests, land and waterways in which indigenous people live, opening them up for industrial development leading to such things as mining, logging and toxic pollution.

IDM are a decolonisation and indigenous sovereignty movement. They articulate resistance to ongoing colonisation, that is, attempts by capital and the Canadian state to appropriate indigenous territory for the purposes of capital accumulation. This is done violently or through various forms of coercion and co-optation such as forms of state-sanctioned recognition that further enable the appropriation of their life-worlds.[39] As part of this struggle, IDM had been at the forefront of climate justice struggles in North America concerning the construction of the 1,700 mile Keystone XL oil pipeline, which was to carry oil from the Athabasca tar sands in Northern Alberta to the United States.[40]

In particular, IDM combined flash mobs and social media – both highly popular forms of activism among young people – with traditional music and dance in a way that bridged generations and cultures, and made some space for building a sense of community. IDM initially used flash mobs to perform indigenous round dances during the pre- and post-Christmas shopping season in 2012 in sites of consumption such as shopping malls to protest against the proposed legislative changes. Round dances are a traditional practice amongst indigenous people in Canada with different ceremonial and spiritual meanings among different First Nations, including healing ceremonies and ceremonies associated with helping the passage of the deceased to the spirit world. They are also used as a form of celebration and as an expression of friendship and unity. The round dance was suppressed in the process of colonisation, and hence round-dance flash mobs represented both a powerful expression of resistance and a practice of cultural regeneration.[41]

Different round dances are examples of indigenous people knowing their place, and consist of circles of people linking hands and singing and dancing to the beat of drums.[42] Round-dance flash mobs have taken place in malls and public spaces across Canada and the United States, including in Saskatchewan, Alberta and Minnesota. Aboriginal activists have gathered at malls and begun beating out a steady rhythm on hand drums to the accompaniment of singing, and subsequently been joined by members of the public.

The use of social media such as Facebook, Twitter and YouTube has extended the reach of IDM's demands (concerning indigenous sovereignty) and tactics from small, spatially dispersed communities to larger communities across Canada. Indigenous communities include registered Indians, non-status Indians, Metis and Inuit. In Canada, young people (under 25 years of age) comprise 46 per cent of the indigenous population, and 45 per cent of registered Indians live on one of the 997 reserves found across the country.[43] In such reserves, dense social ties have enabled relational power to develop amongst a young population who are enthusiastic users of social media. IDM has also extended internationally to the US, Europe and New Zealand.

As wars of words, the round-dance songs articulate deep-seated emotions such as anger, and resentment at the perceived environmental injustice of the proposed legislative changes.[44] Round dances have demonstrated to opponents the grassroots power and continuing strength of indigenous nations. For indigenous participants and viewers, they have promoted cultural pride, connection and experiences of solidarity, and offered for newcomers a welcoming and easy opportunity for involvement. As a powerful visual symbol, flash mobs have carried the resonance of tradition and ceremony while also being fun, loud, entertaining and contagious.[45]

Traditionally performed within indigenous communities, dances and songs have been performed out of place in shopping malls and sites of circulation such as public squares throughout Canada, challenging people's everyday ideas about how these spaces are used. This approach to staying mobile is resistant to police infiltration and pre-emption because of the relational power of friend-to-friend networking involved in organising flash mobs. Finally, I extend this analysis below to show how activists use swarm intelligence to prosecute forms of culture jamming that enable pack and swarm activities.

HACKTIVISM

Hacktivism refers to forms of activism that redesign the use of something, to make something do something it has never done before. It is a form of culture jamming, a practice discussed in detail in Chapter 5. In particular, hacktivism tends to refer to forms of digital activism, or activism that includes a digital component. Recent forms of digital mapping have enabled activists to stay mobile during the changing contexts and conditions of street protests. For example, a map-based smartphone app called Sukey developed in 2011 sought to provide activists with information concerning police containment strategies during protests.[46] The London Metropolitan Police had used indiscriminate containment tactics against student and anti-austerity demonstrations during protests between 2010 and 2012. Such practices penned protestors into 'kettles' – so termed because of the pressure exerted on those within them – in an attempt to contain their mobility and ability to demonstrate. Using OpenStreetMap, Sukey used digital maps of the city in order to visualise police containment practices and designate different types of road (such as those where vehicles did not have a right of way) that were useful for protestors wishing to avoid kettling practices by the police yet intervene in sites of circulation. The app was a collaborative mapping project, drawing upon relational power and the collective knowledge (or swarm intelligence) of protestors, with volunteers uploading data contributions in the form of GPS traces.[47] The app enabled activists to know their place of protest, to anticipate the movement of police deployments, see overviews of actions across the city and to stay mobile by navigating the urban environment through taking shortcuts through alleys, running across green spaces, climbing over walls and so on. Protestors were able to stay informed and create inventive spatial practices in attempts to outmanoeuvre police deployments, frequently resulting in them being, from the perspective of the police, out of place.

Another example of hacktivism was Climate Games, which took place from 29 November to 12 December 2015, as activists from over 150 organisations from around the world converged on Paris to demand a robust, verifiable and legally binding agreement from governments at the twenty-first Conference of the Parties (COP) to the United Nations Framework Convention on Climate Change (UNFCCC) – also known as COP21 – a key terrain of representational power. For activists, the COP21 meeting in Paris represented a classic site of decision in which to intervene: a place where governments would attempt to fashion and sign an international agreement concerning the threats of climate change.

In fashioning their struggle, the coalition of organisations and social movements present in Paris deployed their organisational powers to make claims and demands concerning climate justice, that is, economically and environmentally just responses to climate change. Climate Games merged direct action on the streets with the internet to create a real-time crowd-sourced mapping of climate justice protests in Paris and across the world.

The idea of Climate Games was created by activists from the Laboratory of the Insurrectionary Imagination[48] working with other activists in workshops in various parts of Europe to make some space online by constructing a website and app to inspire and enable activism during COP21. Drawing upon the *Hunger Games* books and films, Climate Games were framed as a contest played between mobile climate justice packs (variously named affinity groups) and their opponents, the police (named 'team blue') and corporations ('team grey').

Activists used an open-source smartphone app available on the Climate Games website[49] to report on their protest actions in real time by anonymously uploading photos, reports and video material onto a digital map (or 'gaming field'). Protestors could also upload information about

Figure 4.1 Stay mobile: climate justice activists swarm the streets of Paris, 2015. Photograph by the author.

the location of the opposition (teams blue and grey) in the global gaming field.[50] Through this digital platform, climate justice activists were able to reframe the site of the UNFCCC meeting in Paris as a 'gaming field' of direct action protests and provide online space for activists to discuss such protests and present oppositional discourses surrounding COP21.

The Climate Games format also provided a decentralised mass alternative to centralised collective organisation. Affinity groups of climate justice activists stayed mobile, as packs of activists undertook over 120 actions in and around Paris during COP21. It also inspired and extended the reach of protest activity far beyond Paris as actions also took place in Australia, North America and Latin America involving coalmine blockades, bank occupations, radio frequency takeovers, the disruption of politician's speeches and so on.[51]

STAY MOBILE

Staying mobile comprises an interplay of movement and fixity (or deter-ritorialisation and territorialisation) and can enable particular places to be claimed, defended, strategically used and/or abandoned, depending upon the strategies and goals of a particular protest. Staying mobile also refers to the necessity that activists continually adapt to changing contexts and conditions, and can enable protesters to evade capture, deploy compositional power in order to make space and keep strategically ahead of their opponents in particular struggles. While the Naxalite movement failed to adapt to changing government tactics and did not fully adapt Maoist ideology and tactics to the placed contexts of rural India, Black Lives Matter, flash mobs and different forms of hacktivism are examples of the success of adaptation and spatial mobility, and of their ability to generate momentum and multi-site communication as crucial mobilising tools. As I have alluded to here, such practices are further enhanced through the language of protest: the slogans, words, demands and online tools that activists use to mobilise others to generate a broader politics of affinity and articulate dissent. It is to this that I now turn.

5
Wage Wars of Words
Testimonies, Communiqués and Culture Jamming

'Wage wars of words' refers to how activists intervene in and create their own media spaces. It refers to the importance of activist representations of events – through words, images, slogans, leaflets, websites, social media and so on – for not only making demands but also for creating alternative ways of thinking about particular issues. The use of words and images is an important strategy in the conflict over representations of events between activists, governments, private corporations and the public.

Embodied acts of resistance are linked in significant ways to what is said and conveyed (through words, images and so forth), and also, therefore, to what communication devices and technologies are used and the role they play in protests.[1] In this way, protests act as forms of media transmitting messages to society, frequently as symbolic challenges that attempt to make dominating power visible and thus negotiable. They pose challenges to the symbolic order of what constitutes permissible thinking and action on specific issues.[2]

In waging wars of words, protestors are usually keen to engage with the mainstream media if they can creatively exploit it to get positive coverage, but they also frequently create their own forms of alternative media in order to present their views. In a world where the contemporary media tends to focus on the sensational and spectacular, activists need to ensure that their actions not only gain public attention but are also framed within a 'battle of the story' – where oppositional ideas and solutions are given 'air time' – rather than as a 'story of the battle' – where the media focuses on the spectacular elements of a particular protest such as property damage or fights between protestors and the police.

In order to influence public perception, protestors need to 'think narratively' by considering how stories and power are interwoven. This requires a consideration of: which audience activists are trying to reach; how a particular issue or conflict is framed; what types of story or imagery are deployed in order to show what is at stake in a particular struggle (rather than tell people what to think about that struggle); and how assumptions or common sense thinking on an issue can be challenged, and in so doing express alternatives and solutions.[3] The use of media enables particular protests to be placed for broader audiences, and also enables a particular protest to extend its reach beyond the place of protest to other locations.

In waging wars of words, activists make some space materially and digitally, through creation of a 'media ecology'[4] that includes technologies such as laptops, USB sticks, memory cards, hand-held digital recorders, cameras, smart phones and so on, as well as physical environments, weather and the availability of shelter and electricity that can impact on the production and transmission of activist messages. A media ecology enables activists to document events, communicate, transmit alerts concerning police locations, surveillance, arrests and violence, and send SMS updates on mobiles and Twitter platforms for information.[5] As I discussed in Chapter 4, such processes can enable activists to stay mobile by adapting and responding to rapidly changing protest events.

The increasing prevalence of the use of such technologies can be traced back to the alter-globalisation protests against the World Trade Organisation (WTO) in Seattle in 1999. Frustrated with the mainstream media's misrepresentation or omission of protests, activists decided to become the media themselves and established their own independent media centre, called Indymedia.[6] Activists set up tents to act as spaces for activists to write reports on protests and upload them along with photos and video material. Indymedia sites subsequently spread to cities in the United States, Latin America, Europe, Asia and Oceania, and were the catalyst for various forms of participatory news publishing and skills-sharing. While these sites were originally relatively static spaces (such as a tent or building), with the advent of smart phones, USBs, wireless and Bluetooth technology, the waging of wars of words have become more mobile and flexible.

Activist media can include stories, testimonials, manifestos, recorded interviews, videos, eye-witness reports, images, the live-streaming of events, and social media and digital platforms. They are the product of and generate relational and compositional power. They create protest and movement cultures and identities, and convey emotions and at times move people to act. The also act as vectors transmitting movement actions, ideas and demands, enable the sharing of information, can educate and sway public opinion, generate solidarity and can create memes – viral frames that allows a story to spread.[7] The use of such communication techniques and various forms of media can enable protests to extend their reach beyond their place of performance, as I will argue in the next chapter. In this chapter I focus primarily on the use of communication and media to articulate the key issues, concerns and frames of particular protests.

Below I consider two examples from the Global South, where the materiality of everyday life is under threat by resource exploitation. In the first, I discuss the power of testimonies and slogans in the anti-dam struggle waged by the Narmada Bachao Andolan in India. In the second, I discuss the use of communiqués and social media by the 'informational guerrillas' of the Zapatista movement in Mexico, resisting the marginalisation of indigenous people. Finally, I discuss forms of protest in the Global North known as 'culture jamming', which target consumer culture using mediated communicative tools.

VOICES OF THE DAMMED IN THE NARMADA VALLEY[8]

While living and conducting solidarity work in 2000 at the resistance camp in the Pawra *adivasi* (indigenous people) hamlet of Domkhedi, in the Indian state of Maharashtra, I attended one of the camp meetings. About 50 of us were huddled under the meeting tent in the hamlet. Under the light of a single hurricane lamp, the faces of *adivasi*s were dimly illuminated. A waning moon cast a silvery light across the Narmada River. In the distance the thrum of a motorboat faded into the darkness. *Adivasi* testimonies filled the space within the meeting long into the night as they recounted the impacts of large dam construction on their valley – the trauma of displacement, the hardships and deceits involved in resettlement, and their ongoing struggle for justice.

The Narmada Valley Development Project (NVDP) entails the construction of a series of dams – 30 mega-dams, 135 medium-sized dams and 3,000 small dams – along the entirety of the 820 mile-long valley of the Narmada River, which flows through the states of Madhya Pradesh, Maharashtra and Gujarat. The project, initiated in 1961, was part of India's post-independence plans to develop its agriculture and industry in an effort to achieve economic and political self-reliance.

The dam-building project is part of official Indian discourses of development that have tended to associate a (Westernised) culture of progress and modernity with development projects such as large dams, and which associates these with the promise of drought alleviation and the provision of sustainable development. Meanwhile, 'non-modern', traditional and indigenous systems of knowledge have been devalued and portrayed as 'unscientific' and 'irrational'. By their very existence, such practices have been perceived by development planners as blocking India's continued modernisation and, more recently, liberalisation. Financing for their construction has come from the Indian state, transnational corporations and international institutions such as the World Bank. In reality, much of the water will be used by commercial agriculture, industry and cities.

However, the construction of the dams along the Narmada River has entailed environmental, economic, cultural and political erasures that cast the entire valley as a site of destruction that will appropriate the resources, territories and life-worlds of peasant farmers and *adviasi*s. Environmentally, the entire river valley will be submerged, destroying forest, farmland and river ecosystems. No thorough environmental impact assessment has been made. Economically, once all of the dams are constructed, an estimated 15 million people will be displaced and their livelihoods destroyed. Government resettlement and rehabilitation policies have a woeful record in India, and in the case of the Narmada dams, state governments have acknowledged that they do not have adequate land to resettle those displaced.[9]

Culturally, the Narmada Valley is home to a range of different people, including wealthy cash-crop farmers of the Nimar region, and *adivasi* subsistence farmers such as the Bhil, Bhilala and Pawra. The valley has been these peoples' home for generations, and the river is of great cultural and spiritual importance to all the valley's communities. Displacement will sever these people's cultural and spiritual connections to their homes and lands. Politically, people's consent for the development project has never been sought. When they have resisted the dams and their enforced displacement, they have been met with arrest, detention, harassment and physical violence perpetrated by state government police forces, frequently in direct violation of India's laws.

The struggle against the dams has been ongoing since 1985, waged by the Narmada Bachao Andolan (NBA), or Save the Narmada Movement. The NBA has waged two interrelated forms of struggle. First, it has waged a representational struggle over the meaning of such processes as democracy and development in India. This has involved a discursive conflict over very different imagined geographies, pitting the spaces of erasure perpetrated by the state and transnational corporations against the lived space of *adivasis* and peasant communities. This war of words has included the testimonies, songs, poems and *naras* (slogans) of the NBA, as well as detailed research and analysis on the impacts of the dams and on sustainable development alternatives. Second, the NBA has waged a material struggle in sites of production, social reproduction and destruction over land and water resources, where the people of the valley have struggled to protect their cash-crop and subsistence livelihoods and their cultures against submergence by the dams. In addition to demonstrations and rallies, the NBA has deployed compositional and organisational power, making some space through establishing *satyagraha* ('truth force') camps wherein villagers in the area near the dams threatened with submergence have resisted eviction and pledged to remain in their homes even at the risk of being drowned.

The Power of Words: Analysis, Testimony, Slogans

The NBA uses demonstrations as forms of public confrontation with officials involved in the Narmada project ranging from local revenue or forestry officials to state leaders and World Bank officials.[10] At these events, the NBA demands that the state be held accountable for actions that belie its claim to protect and enhance the well being of its citizens. Development and other discourses become contested in such sites, drawing upon local experiences to critique state practices. The NBA also intervenes in sites of collaboration by using the established legal channels of the state, such as India's Supreme Court, to contest the legitimacy of development policy in the Narmada Valley.

Discursive resistance, like its material counterpart, disrupts state discourses regarding development. First, the NBA uses and produces academic and activist analyses of the ecological, social and economic effects of the construction of big dams.[11] These analyses are grounded in a deep knowledge of place (concerning the impacts of the dams on the economic and cultural life of the villages and towns along the Narmada, as well as the river's ecosystem), and are deployed at both state and national levels to lend legitimacy to the NBA's dealings with the state (for example, to challenge the construction of the dams in India's Supreme Court). They are also deployed by international lobbying and solidarity groups, enabling the movement to extended its reach[12]

Second, the NBA utilises *adivasi* and peasant testimonials to address state injustices such as encroachment on tribal land, harassment, corruption and violence, as well as the effects of displacement and resettlement upon *adivasi* communities. Such eyewitness accounts are deployed in villages to recruit more supporters to the struggle, and in confrontations with state and national officials. As Kay Warren has argued, testimonials 'personalise the denunciation of state violence', representing 'eye-witness experiences ... of injustice and violence', and 'involve the act of witnesses presenting evidence for judgement in the court of public opinion'.[13]

For example, Bawabhai, an *adivasi* from the dam-threatened village of Jalsindhi, evoked his knowledge of place by explaining to me the desacralisation of place brought about by submergence of villages by the dams:

> Many temples (*mandir*) have been submerged. The blame is on the government as they are responsible for submergence. God exists in everything and therefore it is not right to submerge places where so many people had their faith, for *advisasi*s and non-*advisasi*s. Jalsindhi Mata was a small temple, now it is completely submerged. Now we offer devotion from the village only in the name of our goddess, from the upper bank. They have drowned our gods.[14]

Testimonials also provide justification for action and resistance. The crucial power of these testimonials is to construct a reality where other *adivasi*s will reconsider state efforts at resettlement and re-evaluate the risks and benefits of working with the NBA. Such testimonials are normative, constructing a political climate in which resistance occurs. By narrating corrupt state actions (in the case of development), *adivasi*s and activists produce *adivasi* resistance. As such, testimonials result in mobilisation against the dams, since *adivasi*s speak in the language of other potentially displaced people and recount experiences germane to their existence.[15] Testimonials act to reinsert *adivasi*s and their experiences into a social system that marginalises them and makes them invisible. They symbolise a revolt against invisibility.

Supporting these testimonials are highly symbolic displays of reverence, which commemorate those who have died or who have been injured as a result of state violence. In the summer of 1999, I attended a memorial for Rehmal Vasave (an *adivasi* teenager killed by police),

Figure 5.1   Wage wars of words: NBA poster critiquing
the dams, Narmada Valley, India, 2000. Photograph by
the author.

where the sacrifice of those killed or assaulted by the state was praised
and the righteousness of their cause affirmed. As Dedlibhai told me a
year later in the village of Nimgavhan:

> It's our land, forest, river and life. Not only governments but foreign
> companies are trying to snatch this away. We have taken the challenge
> to save human life, protect human rights. Our Rehmal has sacrificed
> his life for us. We too are ready for that.[16]

By memorialising such incidents, the NBA is able to transform them
into touchstones for further action. The geography of testimonials is
also important. They take place in sites of social reproduction at village
meetings (to inform and mobilise local people) and at sites of decision

during rallies and demonstrations outside or within government offices with the national and international press present. Testimonials offer visceral and passionate denunciations of state practices, and underpin the moral legitimacy of the NBA's struggle. They are entered as grievances in the public record, and together with other testimonials form part of the overarching narrative of *adivasi* experience in the Narmada Valley.

Third, slogans (*naras*) also form an important component of the NBA's discursive repertoire. The term 'slogan' owes its origin to the Scottish Gaelic word *sluagh-gairm* ('army shout'), referring to the war cries and assembly signals of the clans of the Scottish Highlands. In the context of the NBA, *naras* articulate both a sense of injustice and religious and moral legitimacy in order to motivate people to resist.

During 1999 and 2000, I participated in two *satyagraha* camps and several of the NBA's actions. Throughout my engagement with the movement, *naras* acted as the pulse of the NBA. In addition to articulating movement demands at demonstrations and rallies, and on protest banners and political graffiti, *naras* were utilised for a range of purposes within the spaces of meetings and rallies, helping to strengthen the relational power between villagers in the Narmada Valley. *Naras* were used to: lift the energy of a meeting; introduce, punctuate or conclude a speech, and in so doing to incite moments of participation and inclusion amongst the audience as they echoed or amplified the speaker's voice; propagandise the goals and demands of the movement; act as a greeting or farewell when activists arrived at or left a meeting in a village; and acted as a call-and-response mechanism that unified the speaker and the audience.

The *naras* of the NBA also represent the confluence of different, yet braided counter-hegemonic discourses. First, they articulate activists' political intent in the face of threatened submergence and attempted eviction – to take an example, *Koi nahi hatega, bandh nahi banega* ('We shall not move, the dam will not be built'). Second, they articulate *adivasi* knowledge of place – *Jungal jameen kuni chee, amri chee, amri chee* ('To whom does the forest and land belong – it is ours, it is ours'). Third, they articulate how the NBA makes some space through self-organisation within *adivasi* areas of the valley – *Humara gaon, mai humara raj* ('Our rule in our villages'). Fourth, they articulate a critique of the government's economic policy – *Vikas cha me yeh, vinaash nahi* ('We seek right development, not destruction'). Fifth, they articulate movement solidarity amidst cultural differences as the NBA extends its reach – *Hum sub ek hai* ('We are all one'). Finally, they articulate the self-sacrifice inherent in the practices of *satyagraha* – *Doobengeh par hatengeh nahi* ('We shall drown but we shall not move').

*Nara*s run like a river through the NBA's political practice. They echo across the valley as activists call to one another from boats and hills. This discourse forms a river that braids the ideology of the NBA together through its participants and weaves the NBA's different constituencies – *adivasi*, caste Hindu, men, women – together. As a group of us walked over the hills beside the Narmada in the dark of night, NBA leader Medha Patkar put it like this: 'We will mark out the space between ourselves with slogans'.

Through its material and discursive resistance, the NBA has been able to construct itself as a non-violent movement of small peasants and *adivasi*s confronting the destructive development of local and national governments. Their war of words has also been strengthened by powerful images of villagers chest-deep in rising waters caused by monsoon flooding of the dammed river, pledging to drown rather than move from their homes and livelihoods. It has enabled the NBA to extend its reach, conducting its resistance simultaneously across multiple scales.

First, the NBA has grounded its struggle against the dams in the villages along the Narmada Valley, mobilising *adivasi*s, peasants and rich farmers to resist displacement. The NBA has been able to use their local knowledge of place to facilitate communication between disparate communities and to mobilise, at times, tens of thousands of peasants to resist the dams by drawing upon local religious beliefs in the river goddess Narmada and the threats posed to her by the dams.

Second, the NBA has served petitions to the Supreme Court of India, and has established, and participated as a convener in, the National Alliance of People's Movements (NAPM), a coalition of different social movements in India collectively organising to resist the effects of liberalisation on the Indian economy. Third, the NBA has forged international links with various groups in part through the *satyagraha* camps that have enabled the building of relational and organisational power. For example, during the camp that I joined between July and September 2000, participants included local *adivasi*s, farmers from the nearby Nimar Plains, members of NAPM, Indian students, activists from a range of Indian women's, environmental and peasant organisations, and international activists, researchers and students from Britain, Canada, the United States and the Netherlands. In addition, groups such as International Rivers Network, Friends of the Earth, Survival International, the Narmada Action Committee and People's Global Action (see Chapter 6) have engaged in international solidarity. Although dam construction has proceeded, it has been significantly slowed by the struggle of the NBA, which continues to the present.

While slogans and testimonies have played a key role in the NBA's struggle, in Mexico, the Zapatistas have used the motive force of communiqués and a broader range of social media as an integral part of their struggle.

THE ZAPATISTAS AS INFORMATIONAL GUERRILLAS[17]

[The storm] will be born out of the clash between two winds, it will arrive in its own time, the coals on the hearth of history are stoked up and ready to burn. Now the wind from above rules, but the one from below is coming, the storm rises ... so it will be.[18]

On 1 January 1994, media vectors around the world carried the dramatic news that ski-masked guerrillas had captured the town of San Cristobal de las Casas in the Mexican state of Chiapas and declared war on the Mexican state. As the drama unfolded, it became apparent that the EZLN (Ejercito Zapatista Liberacion National), or Zapatistas, as they became known, did not see themselves as the vanguard directing a struggle to seize state power. Rather, they demanded the democratic revitalisation of Mexican civil and political society, and autonomy for, and recognition of, indigenous culture.

Perhaps more than any previous guerrilla insurgency, the Zapatistas consciously sought to use the media as an integral part of their overall political strategy, and used an articulate, humorous and poetic spokesperson in the masked figure of Subcommandante Marcos.[19] The Zapatistas' war of words articulated challenges to both the material political and economic power of the Mexican state, and to the monopoly of representations – imposed by political elites – of Mexico as an 'emerging market', economically and politically stable for foreign investment. In so doing, they represented themselves so as to be seen and heard by national and international audiences through various forms of media.

The Zapatistas had developed relational and organisational power among indigenous Mayan peasant communities over a decade before they staged their uprising in a spectacular manner to ensure maximum media coverage, and thus gain the attention of a variety of audiences – including civil society, the state, the national and international media and international finance markets. Their war of words consisted of three interwoven facets: the physical occupation of space; the mediation of images; and the discursive power of communiqués. What differentiates the Zapatistas, in part, from many of the earlier political struggles that have peppered the history of Latin America is not so much their waging of guerrilla war on both the landscape and the mediascape, but rather the strategic importance of images and words in their struggle.

Material Struggle, Images of Resistance

The initial uprising in Chiapas articulated indigenous and *campesino* (farmer) resistance to the North American Free Trade Agreement (NAFTA) between Mexico, the United States and Canada. Such trade agreements have been key platforms for the advancement of neoliberal privatisation policies around the world. For example, NAFTA would allow the United States to flood the Mexican market with US subsidised corn and undercut the market for Mexican corn, threatening farmers' and indigenous people's livelihoods. It also necessitated the reform of Article 27 of the Mexican constitution, opening up communally owned smallholdings for sale to national and foreign capitalist interests. The Zapatistas deployed organisational power and made some space by occupying San Cristobal de las Casas and other Chiapan towns. The initial occupation of San Cristobal was a masterstroke of public relations, serving to awaken the Mexican government and the international media to what at first appeared to be a traditional guerrilla insurrection. While the EZLN communicated their declaration of war to the Mexican government,[20] they also attacked symbolic targets. The Zapatistas mobilised between 1,200 and 1,500 fighters for the New Year's Day offensive (400 in San Cristobal, 300 in Ocosingo). They attacked and damaged sites of decision such as town halls, police stations and the municipal palace of Altamirano, before disappearing back into the jungle.[21]

However, the national and international media conveyed images of ski-masked guerrillas engaged in what appeared to be the creation of a liberated zone. The guerrillas were out of place in the dominant script crafted by the Mexican government and its NAFTA allies, in which Mexico was represented as a newly emerging market ripe for foreign investment. Their appearance targeted a key site of assumption of the NAFTA signatories, namely that their free trade policies would proceed without hindrance.

In reality, the Zapatista 'liberation' was brief, lasting 30 hours, before the Zapatistas melted back into the Lacandon jungle. They were more concerned with making some space in virtual networks than permanently securing control of Chiapas's major towns. Their material occupation of towns was symbolic, staged to gain access to media vectors. This focused upon the images of a guerrilla war rather than the undertaking of a protracted armed struggle.

However the Zapatistas *are* a guerrilla force that has engaged in several armed conflicts with the military. Their tactics of armed struggle were developed over many years and drew upon the Zapatista's knowledge of place:

> We learned our tactics from Mexican history itself ... from resistance to the Yankee invasion of 1846–1847, and from popular resistance to the French intervention, from the heroic deeds of Villa and Zapata and from the long history of indigenous resistance in our country.[22]

The Zapatistas made some space for themselves in the Lacandon jungle that served as a material base for the creation, organisation and subsequent emergence of their resistance. The Lacandon jungle was a site of production (for growing food) and a site of social reproduction (as guerrillas and their families conducted everyday tasks of sustaining the rebel population). It provided a material sanctuary for the guerrillas when they were faced with the military might of the Mexican army and air force, and a base from which they waged their war of words.

Throughout the conflict, the EZLN have also stayed mobile, emerging sporadically from the jungle, taking over small villages and towns, and then retreating back into the jungle when the army has appeared. In addition, the Zapatistas have located their general command in mobile encampments. The ability to stay mobile has frustrated the government in its attempts to curtail the insurgency, although the Zapatista rebellion has presented little military threat to the Mexican state. However, ski-masked, armed guerrillas occupying Chiapan towns, and then 'disappearing' back into the jungle, was a visually powerful image with a deep cultural and political resonance throughout the Americas.

The tactic of armed struggle and the discourse of insurgency has an important place within both Mexican society – whose revolution was secured by force of arms – and within a region that has witnessed numerous armed rebellions, including the Cuban and Nicaraguan revolutions, and armed insurgencies in El Salvador, Columbia and Guatemala. Through knowing the place of rebel insurgency in the popular political imaginary, the Zapatistas were able to be seen and heard by a variety of 'publics', including the Mexican government, national and foreign investors, as well as Mexican and international civil society. The Zapatistas have used a variety of media – including the internet, newsprint media and radio – to give voice to their struggle and demands, particularly through the use of communiqués.

Communiqués

Out of the mouths of the guns of the faceless men and women spoke the voices of the landless campesinos, the agricultural workers, the small farmers, and the indigenous Mexicans. The voice of those who have nothing and deserve everything.[23]

The eloquence of Zapatista communiqués was partly responsible for their success in extending their reach. They helped to develop relational power by mobilising civil society to put pressure on the federal government to end its early military attempts to destroy the EZLN. They also helped to create deep sympathy for the Zapatistas throughout Mexico. Several US publications have published interviews with Sub-commandante Marcos (including the *New York Times* and *Vanity Fair*), and these helped contribute to international public opinion calling for a peaceful settlement in Chiapas.

The EZLN's communiqués were cast in different voices. Many of the early ones used diplomatic language as the EZLN struggled to be formally recognised as a belligerent force by the federal government, and to denote the official nature of the EZLN organisation. Meanwhile, many of Marcos's 'personal' letters were written in the idioms of Mexico City's streets and universities. Marcos used many voices: self-mocking and wisecracking; those of a letter writer, storyteller and poet; the voice of a visionary community, of the indigenous people of Chiapas, and as the voice of the EZLN's Comite Clandestino Revolucionario Indigena-Coordinadora General (CCRI-CG).

The communiqués articulated the amalgamated injustices, grievances and demands of the peasants and indigenous people of Chiapas. Such demands had little to do with ideology and everything to do with peasants' needs, and hence their resonance with civil society. In addition to the print media, EZLN communiqués were also posted on numerous Mexican news bulletin boards on the internet, and then downloaded and photocopied by groups in other countries (such as Spain) to be handed out at demonstrations in solidarity with the Zapatista cause.

Nevertheless, the materiality of armed struggle within the Zapatista rebellion is an integral part of their strategy, and one without which the political effect of the Zapatistas' war of words would have been negligible. As Marcos noted: 'We did not take up arms to appear in the newspaper. We took up arms so as not to die of hunger'.[24]

Targeting a site of assumption, the Zapatistas articulated a counter-discourse to that of the Mexican government, which had attempted to create the appearance of Mexico as a newly emerging market, stable for foreign investment and an economically viable partner within NAFTA.[25] This counter-discourse served three ends. First, to make people aware of the unequal distribution of land and economic and political power in Chiapas; second, to challenge the neoliberal economic policies of the Mexican government and articulate the long-term detrimental effects of NAFTA upon the peasant economy; and third, to articulate a call for the democratisation of civil society, which in turn enabled the creation of a political space in which numerous indigenous and peasant organisations could articulate their own political and material challenges. Part of this discourse challenged the invisibility of indigenous people in Mexico, viewed as unworthy of consideration in economic decision-making. The effect of the Zapatistas has also been to cause capital flight from Mexico at the start of NAFTA, and the devaluation of the Mexican peso.

Autonomous Spaces

Perhaps the most important outcome of their struggle is that the Zapatistas have deployed compositional and organisational power, and made some space by practising democratic autonomous self-management within indigenous communities concerning the provision of educational and health projects, making decisions through community-based assemblies.

For example, Zapatista autonomous municipalities have been established within Chiapas, where communities govern themselves and participate in conflict resolution between communities through practices of autonomous justice. They have established sites of social reproduction by constructing their own autonomous primary schools and health clinics – where they practice a combination of traditional Mayan medicine and allopathic care. They also engage in autonomous subsistence through traditional farming and agro-ecological techniques.[26] A network of community-controlled radio stations broadcast in indigenous languages (such as Ch'ol, Tojolabal, Tzeltal and Tzotzil), while their schools teach in both Spanish and indigenous languages.[27] Through these practices, the Zapatistas articulate an alternative to *indigenismo* – the politics of assimilation of indigenous peoples into the fabric of the Mexican state – by advocating the politics of *indianismo*, which articulates the indigenous world view and promotes Indian political autonomy.[28]

Importantly, the Zapatistas instituted the Women's Revolutionary Law in 1994 – the first guerrilla movement to include demands for women's rights in its struggle. Indeed, it was women activists who led the occupation of San Cristobal de las Casas, and the female Zapatista Commandante Ramona was at the centre of subsequent peace dialogues. The Women's Revolutionary Law ensured women rights to autonomy; rights to participate in revolutionary struggle, community affairs and positions of authority; rights to health care and education; and rights to sexual and reproductive freedom and freedom from physical assault. As a result, the position of women in Zapatista-controlled territories has greatly improved. However, all politics is a process, and women admit that they have not yet accomplished all of these rights fully.[29]

In extending their reach, the Zapatistas have inspired activists around the world. In particular, they have established *encuentros* ('knowledge encounters') where a diversity of visions, responses and solutions to neoliberal capitalism have been articulated, summed up in the Zapatista notion of 'one world in which many worlds fit'. Following a Zapatista *encuentro* in 1996, where Subcommandante Marcos called for an international network of resistance against neoliberalism, alter-globalisation mobilisations began to emerge around the world, inspired by the Zapatista call. Two particular initiatives emerged from these mobilisations, the international direct action network People's Global Action (see Chapter 6) and the World Social Forum (WSF), a conference of social movements from around the world that first met in 2001 in Porto Alegre, Brazil. The WSF has subsequently been held in various parts of Latin America, Asia and Africa, and has also generated national and regional social forums around the world, including in North America and Europe.

From 2006 onwards, the Zapatistas organised the Other Campaign, with Subcommandante Marcos travelling across Mexico to meet with other indigenous communities, NGOs, labour and peasant unions, students, environmental activists, women's groups and so on in order to build support across civil society and agitate for greater indigenous rights and better protection of indigenous peoples under the country's constitution. Over the past 22 years, the Zapatistas have continued to build up and defend their autonomous communities, and have significantly reduced their engagement with the media. In 2014, Sub-commandante Marcos stepped down as spokesperson for the Zapatistas to be replaced by another identity, Subcommandante Galeano. However, the movement's war of words has not ceased. A book of contemporary Zapatista analyses of capitalism and the importance of organisational power was published in 2016.[30]

On New Year's Day 2016, a Zapatista communiqué commemorating the '22nd Anniversary of the War against Oblivion' appeared on digital vectors arguing for ongoing struggle:

> With the death of our people, with our blood, we shook the stupor of a world resigned to defeat.
>
> It was not only words. The blood of our fallen *companeros* in these 22 years was added to the blood from the preceding years ... decades and centuries.
>
> We had to choose then and we chose life.
>
> That is why, both then and now, in order to live, we die.[31]

Since the Zapatistas in particular, activists have increasingly combined technological know-how with political acumen to create sophisticated wars of words concerning a variety of issues. These are frequently combined with material protests to deepen and diversify activist campaigns. A particular form of protest that has become important over the past 20 years is 'culture jamming'.

CULTURE JAMMING

Culture jamming refers to a repertoire of actions and practices attuned to consumerist culture and mass-mediated images and messages that enable activists to intervene in sites of assumption and consumption. Early forms of culture jamming consisted of what was termed 'subvertising' by the

Canadian anti-consumerist organisation Adbusters Media Foundation.[32] This relied on the Situationist International concept of *detournement*, or 'turning around' – interrupting, reusing and reassembling the messages received during everyday consumerist experience (through advertising, for example).

The purpose was to target sites of assumption to reveal the underlying (corporate, neoliberal) ideologies of advertising, mainstream media and political messages, as well as cultural artefacts, and in so doing communicate meanings that were at variance with their original intention. A celebrated example of this was the jammed version of the Stars and Stripes, US national flag, with the stars replaced with corporate logos, conveying the corporate domination of US political and economic life.[33]

A recent expression of *detournement* has taken the form of 'brandalism'. Brandalism is an anti-advertising movement started in London in 2012 that has replaced outdoor advertisements with posters that aim to create awareness on particular issues. For example, at the 2015 climate justice mobilisations in Paris, prompted by the meeting of the signatories to the United Nations Framework Convention on Climate Change, 82 artists from 19 countries created posters using computer graphics that mimicked official advertisements in design and appearance, and replaced 600 public advertisements (on billboards, at bus stops and so on) with artworks that revealed the connections between advertising, consumerism and climate change. One poster showed an Air France hostess holding her index finger to her mouth in the 'keep quiet' signal. The post read: 'Tackling climate change? Of course not, we're an airline. Air France, part of the problem'.[34]

Culture jamming deploys memes that stimulate visual, verbal, musical or behavioural associations and play on the emotions of viewers and bystanders, generating feelings that are out of place (such as humour, shock, shame and anger) in order to catalyse thinking on particular issues. Street art, guerrilla texts, guerrilla theatre (see Chapter 7) and a wide range of communication tactics open up discursive spaces that force people out of their comfort zone, and in so doing open up possibilities for cultural change to be imagined and pursued. However, the oppositional meanings of culture jamming can be incorporated into corporations' own advertising, recommoditizing them. For example, a choreographed flash mob at Liverpool Street Underground station in which hundreds of people broke into a dance medley has received more than 36 million YouTube views. However, at the end of the culture jam, T-Mobile's corporate jingle plays over the company's slogan 'Life's for sharing'.[35]

Nevertheless, culture jamming can be highly effective. For example the Yes Men (led by two activists who are known by the aliases Mike Bonanno and Andy Bichlbaum) exploit cultural codes: the HTML code that composes websites, the behavioural codes that regulate the conduct of corporate media, business and government, and the graphic and linguistic codes that enable appearance in the mass media. They have hacked the WTO website, re-routing the contact e-mail address for information to themselves. As a result, they received invitations directed to the WTO, whereupon the Yes Men posed as WTO representatives and subverted the neoliberal logic of the organisation by delivering outlandish presentations, such as equating slavery with economically sound business practice. In 2004, Andy Bichlbaum posed as a representative of Dow Chemical on the BBC and announced that the company had taken full responsibility for its role in the 1984 gas disaster in Bhopal, India, that had resulted in thousands of deaths, and was willing to compensate the victims. While the event attracted international media attention and resulted in Dow suffering a loss of 2 billion dollars on the German stock exchange, some Bhopal victims believed, for a while, that they were to be compensated.[36]

Other forms of culture jamming such as urban mapping have enabled activists to take the initiative in waging struggle through imaging (see also Chapter 4). For example, a map of the square mile of the City of London was used during protests against the meeting of the G20 in London during 2009. Prior to the protests, activists posted a map online, entitled 'Squaring up to the Square Mile'. The map depicted the City of London with the locations of 47 assorted law firms, banks, carbon trading businesses, energy and other corporation offices, arms trade businesses and financial associations. These sites represented potential targets for direct action protests, although the police had no foreknowledge of which sites would be targeted. The online image of the map established a site of potential by articulating an oppositional challenge to a range of neoliberal enterprises. The mainstream media reported that the City of London faced the threat of being overrun by protestors, while the police described the threat as 'unprecedented'.[37] Therefore, even before the start of the G20 meeting, activists had taken an important initiative in the war of words and images over the role of the City of London as a key engine in driving forward neoliberalism, by generating unpredictability concerning which sites in the Square Mile would be the target of protests. As a result, many businesses shut their offices during the G20 meeting. The use of an image of a map had enabled activists to have a spatial impact on the City of London, and on the discourses surrounding the G20 meeting even before it had begun.

WAGE WARS OF WORDS

Wars of words act as a form of 'semiotic resistance' that articulates the 'desire of the subordinate to exert control over the meanings of their lives, a control typically denied them'.[38] It enables protestors to be seen and heard. The creative utilisation of activist media can create protest cultures, convey emotions that move people to act, and disseminate movement actions, ideas and demands; activist media can also be used to share information, educate and generate solidarity and create memes. Activist representations of events enable the making of demands, and the creation of alternative ways of thinking about particular issues.

The emergence and increasing use of social media has led some to argue that there has been a shift from collective action rooted in mass media to connective action rooted in social networking, what Bennett and Segerberg term 'crowd-enabled connective action',[39] where large-scale mobilisations take place without centralised organisational structures, such as occurred with the Occupy protests. However, networked political actions by more traditional social movements remain critically important. They require embodied face-to-face meetings and interactions in order to develop relational power as a prelude to fashioning organisational power both in placed struggles (such as the NBA and Zapatistas), culture jams such as brandalism and in attempts by protestors to extend their reach, an issue to which I now turn.

6

Extend Your Reach

Convergences, Conferences and Caravans

This intercontinental network of resistance, recognising differences and acknowledging similarities, will search to find itself with other resistances around the world ... [W]e are the network, all of us who resist.[1]

'Extend your reach' refers to how, as protests around the world become increasingly connected through the use of social media such as the internet, activists craft and sustain networks of solidarity at local, national and international scales. This chapter examines the fashioning of embodied forms of activist connection, wherein communicative technologies – such as websites, e-mail forums, social media and digital platforms – are utilised to enable face-to-face, material practices of building solidarity. This is because relational power between activists – such as relations of friendship and trust, characteristics of reputation and charisma – are primarily fashioned through political activity grounded in fact-to-face interaction rather than technologies of communication.[2]

Extending your reach implies a relational understanding of space – that places are the product of greater and lesser flows of people, capital, technology, cultural artefacts, commodities and media – and spatial strategy (as discussed in Chapter 1). The spread of digital media means that those physically present in a particular struggle are invariably part of broader material and digital networks of support and organisation.[3] For protestors located in a particular struggle and wishing to project that struggle beyond their local places of protest, 'the local must be recast outside of itself in order to be established as local'.[4] In other words, the projection of a particular struggle across national and international space helps to define the identity of that place as a site of struggle. However, successful international alliances have to negotiate between action that is deeply embedded in place – that is, local experiences, social relations and power conditions[5] – and action that facilitates more trans-national coalitions.

A useful way to conceive of such processes is through the concept of 'convergence space'.[6] Convergence spaces are composed of place-based movements and groups involved in extending their reach, who together with others articulate certain collective visions (such as unifying values, organisational principles and positions), which generate sufficient common ground to generate a politics of mutual solidarity between participating movements. Convergence spaces involve a practical relational politics of solidarity between movements, bound up in five forms of interaction and facilitation: communication, information sharing, solidarity actions, network coordination and resource mobilisation.

Convergence spaces facilitate spatially extensive political action by particular locally based social movements in order to develop transnational networks of support as an operational strategy for the defence of their place(s).[7] Interactions within virtual space act as a communicative and coordinating thread that weaves different place-based struggles together. These connections are grounded in place- and face-to-face-based moments of articulation such as conferences and international protests. However, owing to differential access to (financial, temporal) resources and network flows, differential material and discursive power relations exist within and between participant movements in such networks. As a result, processes of facilitation and interaction tend to be uneven.

Convergence spaces are also characterised by a range of both horizontal and vertical (or hierarchical) operational logics, and are sites of contested and entangled social and power relations.[8] For example, different groups articulate a variety of potentially conflicting goals (concerning the forms of social change), ideologies (such as those concerning gender, class and ethnicity) and strategies (for example, institutional/legal and extra-institutional/illegal forms of protest). Finally, convergence spaces require particular activists who act as 'networking vectors' responsible for developing and extending the processes of communication, information sharing, interaction and organisation within a network. These activists are often dominant within networks, due to their control of key political, economic and technological resources.

Drawing upon the notion of convergence space, I will discuss two recent examples of extending one's reach, both drawn from my scholar-activist activities. As in Chapter 5, I consider forms of communication and the use of media, but here I ground their use in attempts by protestors to spatially extend their struggles to find common cause with others. In the first example, I examine People's Global Action (PGA), the alter-globalisation network that attempted to fashion an international network of social movements and activists against neoliberal capitalism. In the second, I examine one of the constituents of PGA – a farmers' movement in Bangladesh – and their efforts after the demise of the PGA network to deepen solidarity and cooperation with other famers' movements in their region.

CONVERGENCE SPACE 1: PEOPLE'S GLOBAL ACTION[9]

From the mid 1990s onwards, emerging terrains of international resistance to neoliberal globalisation began to emerge, as different social movements representing different terrains of struggle (such as trade unionists, environmentalists and indigenous peoples' movements) experienced the negative consequences of neoliberalism.[10] What became termed 'alter-globalisation' was a movement of marginalised groups and social movements at local and national scales that attempted to forge wider alliances in protest against their growing exclusion from global neoliberal economic decision-making, and posed alternative practices grounded in economic, social and environmental justice.[11] A key network in this process was PGA.

The PGA network owed its genesis to the international encounter between activists and intellectuals that was organised by the Zapatistas in Chiapas in 1996. At the encounter, the Zapatista leader Subcommandante Marcos declared that those present should construct an intercontinental network of resistance against neoliberalism (as quoted at the start of this chapter). In Spain the following year, the idea of a network that brought together different resistance formations was launched by ten social movements including the Movimento dos Trabalhadores Rurais Sem Terra (MST, Landless Workers Movement) of Brazil and the Karnataka State Farmers Union (KRRS) of India. The official 'birth' of the PGA was February 1998, and its purpose was to facilitate the sharing of information between grassroots social movements. The PGA organised an alternative conference at the 1998 ministerial conference of the World Trade Organisation (WTO) in Geneva between social movements from across the majority world, including the MST, KRRS, the Movement for the Survival of the Ogoni People (Nigeria), the Peasant Movement

(Philippines), the Central Sandinista de Trabajadores (Nicaragua) and the Indigenous Women's Network (North America and the Pacific). Together they called for resistance to neoliberal globalisation.

The PGA represented an international network for communication and coordination between diverse social movements, whose membership cut across differences of gender, ethnicity, language, nationality, age, class and caste. It was a unique attempt to fashion solidarity between peasant farmers, trade unions, autonomist groups and other activists. The broad objectives of the network were to offer an instrument for coordination and mutual support at the global level to those resisting corporate rule and the neoliberal capitalist development paradigm. It also sought to provide international projection of their struggles, and aimed to inspire people to resist corporate domination through civil disobedience and people-oriented constructive actions. The PGA also established regional networks – for example, PGA Latin America, PGA Europe, PGA North America and PGA Asia – to decentralise the everyday workings of the network. I will focus on PGA Asia as I was involved in facilitating this network between 2001 and 2006.

The principal means of regular communication across geographic space was through the internet (the PGA established its own website and a dedicated e-mail list).[12] The principal means of materialising the network was through activist conferences, activist caravans and global days of action that I will discuss below. In the PGA network, there was a general rejection of organisational models based on verticality and hierarchy. Instead, the PGA deployed organisational power through 'horizontal coordination', which reflected a decentred networking logic.[13] Such transnational networking processes generated an open communicative infrastructure for the production and circulation of oppositional identities, discourses and practices.

The network articulated certain unifying values – or collective visions – to provide common ground for movements from which to coordinate collective struggles. These were:

A very clear rejection of capitalism, imperialism and feudalism; and all trade agreements, institutions and governments that promote destructive globalisation.

We reject all forms and systems of domination and discrimination including, but not limited to, patriarchy, racism and religious fundamentalism of all creeds. We embrace the full dignity of all human beings.

A confrontational attitude, since we do not think that lobbying can have a major impact in such biased and undemocratic organisations, in which transnational capital is the only real policy-maker.

A call to direct action and civil disobedience, support for social movements' struggles, advocating forms of resistance which maximise respect for life and oppressed peoples' rights, as well as the construction of local alternatives to global capitalism.

An organisational philosophy based on decentralisation and autonomy.[14]

Although PGA Asia was a grassroots-based, decentralised network, it nevertheless involved both newer and more traditional political formations, including NGOs, unions and leftist parties. PGA Asia operated as a network wherein participant place-based movements tended to be organised through more vertical structures and logics: hierarchy, elections, delegation and, in some cases, political party structures too. For example, among the movements involved in PGA Asia were the Bangladesh Krishok Federation (BKF), which held internal elections for a series of hierarchical functional positions within the movement; the Karnataka State Farmer's Union (KRRS) from India, which participated in electoral politics in order to draw attention to rural, grassroots issues; the All Nepal Peasants Association (ANPA), which operated similarly, and was also affiliated with the Communist Party of Nepal (Unified Marxist-Leninist); and the Assembly of the Poor, Thailand (AoP), which comprised of a network of anti-dam, peasant, student and labour movements, each with their own differing modes of operation.

PGA Asia was concerned with five principal processes of facilitation and interaction between participant movements in order to deepen relational power. It acted as a facilitating space for communication (for example, e-mail, web-sites, newsletters, telephone, fax, and face-to-face meetings such as conferences); information-sharing (for example, concerning the effectiveness of particular tactics and strategies, knowledge of place-specific legal issues and local geographies); solidarity (for example, demonstrations of support for particular struggles such as protests); coordination (for example, organising conferences, meetings and collective protests); and resource mobilisation (for example, of people, finances and skills). All of these processes sought to extend the reach of participating movements' struggles through forging solidarity and enhancing struggles.

The sustainability of these processes over time required the development of strong interpersonal ties between participant movement activists that provided the basis for the construction of collective identities.[15] This, and the coordination of joint actions across space (such as global days of action against particular neoliberal institutions such as the WTO), while in part effected through communication via the internet, also required face-to-face processes of communication. These facilitated the exchange of experiences and ideas between activists and enabled strategies to be developed in secure, sequestered sites beyond the surveillance that accompanies any communicative technology in the public realm. Moreover, they provided an opportunity, beyond mass actions, for social movement networks to come together, represent themselves to themselves and others, generate emotional energy and fashion as well as broadcast oppositional discourses.

One of the most spectacular ways that the PGA network effected multi-scalar political action was through global days of action. These are political initiatives whereby different social movements and resistance groups from around the world coordinate around a particular issue or event in a particular place. By identifying structures of power within the global political field, social movements liaised with others to establish common targets of protest that symbolised global economic and political power such as the World Bank and International Monetary Fund (IMF).[16]

For example, the PGA was partly responsible for putting out calls for, organising and participating in a series of global days of action that emerged as one of the most potent forms of alter-globalisation protest. These took place in sites of decision, such as those cities where the key architects of neoliberalism – for example, the G8,[17] World Bank, IMF and WTO – were due to hold their business meetings. Further, the places where such global days of action were held – such as Seattle in 1999 (when the WTO meeting was successfully halted), Prague in 2000 (when the World Bank and IMF meeting was disrupted), Genoa in 2001 (where people demonstrated at a G8 meeting) and Barcelona in 2002 (when demonstrations occurred alongside a European Union meeting) – became critical articulations of the workings of networks such as the PGA.

Not only were there ongoing demonstrations and various forms of direct action aimed at disrupting such meetings such as targeting sites of circulation, but activists from various struggles deployed compositional power and created their own meeting spaces. Activists made some space at these protests through the construction of activist infrastructures as sites of social reproduction (including sleeping spaces, kitchens, media sites and legal support) to sustain their activism during the protests. Local activists had to know their places in order to be able to establish activist infrastructures, and plan demonstration routes and protest sites. In addition, activist counter-forums waged wars of words against their opponents, critiquing the social, economic and environmental injustices associated with neoliberalism. These spaces were sites of assumption and sites of potential in that neoliberal orthodoxies were challenged and economic, political and environmental alternatives were discussed and broadcast to the international media. PGA activists collaborated with other networks in these protests, and were also able to hold network meetings to discuss future PGA activities.

For the PGA, one of the most effective means of sustaining the network, enabling spatial coordination and building relational and organisational power between activists was through international and regional conferences and meetings that provided material spaces within which representatives of participant movements could converge and discuss issues that pertained to the functioning of the network. The purpose of such conferences was to develop solidarity between different social movements in the form of interpersonal communication, exchange of information, coordination of actions, mutual support, and the mobilisation of collective resources. Hence, international conferences were held in Geneva (1998), Bangalore (1999) and Cochabamba (2001), while regional meetings were held in Latin America (2000, 2003), South Asia (2000, 2004), Europe (2001, 2002, 2004, 2006) and North America (2001).

Specific symbolic sites were chosen for the location of PGA international conferences. The PGA conferences in Bangalore and Cochabamba were chosen partly because they had been the sites of successful resistance by popular mass movements against transnational corporations pursuing a neoliberal agenda. In the case of Bangalore, Monsanto and Cargill had both faced successful opposition to their attempts to introduce genetically modified cotton seeds and field trials in the Indian state of Karnataka, by the KRRS – the host of the PGA conference. In Cochabamba, Bechtel faced successful opposition to their attempts to privatise the city's water supply by a popular coalition of students, business people, labour unions and peasants. This included the Six Federations of the Tropics (coca farmers) who jointly hosted the PGA conference with the National Federation of Domestic Workers.

When social movements acted as hosts for PGA international or regional conferences, their struggles were given a certain amount of national and international projection (for example, through the media) as a result. Moreover, the grassroots members of a movement received a boost in morale when activists from around the world visited, and articulated support for, their struggles. According to the activists who attended them whom I met with both during and after some of these events, PGA regional and international conferences enabled grassroots activists to: learn about other struggles in other countries and decrease their sense of isolation; communicate with other activists from other countries; share tactics and strategies; and generate a sense of solidarity between movements. For example, at the 2004 PGA Asia conference held in Dhaka, which I helped to organise, a Malaysian activist from the Borneo Indigenous Peoples' and Peasants' Union explained to me the importance of sharing experiences of activism:

> We wanted to share our experiences of struggle. We don't have many linkages to other movements or the space to speak. The Dhaka conference provided us with that opportunity and the space to speak.[18]

Such communication enabled the creation of common ground between movements.

What is important about solidarity-building in such convergence spaces is that activists' identities are mutable rather than fixed.[19] Solidarity, as Antonio Gramsci argued, is the ability to be transformed, to be open to others in the process of fashioning common ground, as a precursor to mutual solidarity. The point was explained to me by a Nepali activist:

> Dhaka provided a forum to share our work and experiences with others in different parts of Asia, others who have similar problems to us. We were able to share political views and to identify our common ground. The local and the national are not enough because globalisation has intensified the exploitation process across the world. We need to develop global solidarity.[20]

Another key practice by which the PGA facilitated solidarity-building was through activist caravans.[21] PGA caravans were organised in order for activists from different struggles and countries to communicate with one another, exchange information, share experiences and tactics, participate in various solidarity demonstrations, rallies and direct actions, and attempt to draw new movements into the convergence. Caravans enabled representatives from overseas struggles to stay mobile, visiting many different locally placed struggles within a country.

The emphasis on such processes was the two-way communication regarding struggles, strategies, visions of society and the construction of economic and political alternatives to neoliberalism. The caravans included an intercontinental caravan in 1999, which brought 450 representatives of farmers' movements from South Asia to twelve European countries in order to meet activists from European movements and conduct actions, a US caravan that culminated in the WTO protests in Seattle in 1999, and caravans before and after the PGA conferences in India and Bolivia.

Concerning organisational power, PGA Asia was facilitated by social movements within the network that acted as regional and sub-regional convenors. However, much of the organisational work – preparing, organising and participating in discussions, meetings, conferences and caravans – was conducted by key activists and movement contacts (usually movement leaders or general secretaries) who helped to organise conferences, mobilise resources (such as funds) and facilitate communication and information flows between movements and between movement offices and grassroots communities. The 'networking vectors' that possessed English language skills (the lingua franca of PGA Asia) and were computer literate constituted the 'imagineers' of the network. They conducted ideational labour, attempting to 'ground' the concept or imaginary of the network (what it was, how it worked, what it was attempting to achieve) within the grassroots communities who comprised the membership of the participating movements.

Because of their resource access and skills such as communication, experience of activism and meeting facilitation, the imagineers tended to wield disproportionate power and influence within the network. Able to stay mobile (both physically – in that they had the time and resources to travel outside of their home countries – and through their access to distance-shrinking technologies), they performed much of the routine work that sustained the network. They possessed the cultural capital of (usually) higher education, and the social capital inherent in their trans-national connections and access to resources and knowledge.[22] Within political networks, such groups and individuals not only routed more than their 'fair share' of informational traffic, but often determined the 'content' of that traffic. They did not (necessarily) constitute themselves out of a malicious will-to-power; rather, relational power defaulted to them through the characteristics noted above and personal qualities like energy, commitment and charisma, and the ability to synthesise politically important social moments into identifiable ideas and forms.[23]

However, because of the importance of such imagineers in network operations, there was a degree of grassroots alienation with the PGA Asia network. For one Thai activist, the operational logic of the network was underpinned by 'literate' and conceptual communicational forms (such as the writing of e-mails and documents, the analysis of how networks function), whereas the operational logic of most grassroots movements was based upon oral communication, a point she made while discussing the PGA with me in Bangkok after the Dhaka conference:

> There is a real limitation to the capacity of grassroots movements to take ownership of the process. Movements do not know each other very well, and some South East Asia movements do not really know the PGA process at all. Thus participation is limited and language affects this too. Most movements are based on oral communication, whereas the PGA process is more literate and concept-based, thus it is difficult for grassroots movements to understand.[24]

Other activists articulated the need for more traditional, tangible (verticalist) organisational structures than the (horizontal) notion of a 'coordination tool' implied. The PGA imaginary remained abstract to many grassroots activists, for whom the networking logic of many direct action groups (and the PGA) was unfamiliar, as one BKF activist remarked during a meeting in Dhaka: '[We] have to disseminate information to people in rural areas, but so far they have not been able to visualise what the network is'.[25]

As an attempt for social movements to collectively extend their reach, and in so doing develop solidarity with others in order to strengthen their own struggles, the PGA was confronted with a range of networking problems that, in part, were a symptom of its global vision. The network comprised a diversity of movements with different organisational philosophies, ideologies, goals and cultural beliefs. Certain voices tended to dominate discussions and planning, particularly movement leaders and the imagineers. Many of the social movements in PGA Asia had leaderships dominated by men, and activists who evinced patriarchal attitudes and actions. As a result, unequal gender relations made PGA Asia a convergence space where certain dominating powers were entangled with those of resistance. Nevertheless, important connections were made between activists across national borders, and although the PGA had ceased to operate by 2006 – not least because many imagineers left the network to pursue other activities – many of the social movements in South and South-East Asia have continued to work together through other networks. As I discuss below, they have drawn upon the experience of PGA actions to fashion other spaces of activity, cooperation and solidarity.

CONVERGENCE SPACE 2: THE CLIMATE CHANGE, GENDER AND FOOD SOVEREIGNTY CARAVAN[26]

As the morning mist began to clear, we arrived in Bangosonarhat village, Kurigram district, in northern Bangladesh, having travelled all night from Dhaka. Approximately 500 villagers – mostly peasant farmers – gathered to hear speakers from the various social movements that were participating in the meeting. A female Indian activist approached the microphone: 'We are farmer's movements from India, Bangladesh, Nepal, Sri Lanka and Pakistan. On this caravan, we need to build our solidarity, build our strength, and build our alternatives to corporate agriculture.'[27]

This meeting formed one articulated moment of the Climate Change, Gender and Food Sovereignty Caravan (hereafter, 'Climate Caravan') that travelled through Bangladesh in November 2011. Marshalling relational, compositional and organisational power, two of PGA Asia participating movements, the Bangladesh Krishok Federation (BKF) together with the Bangladesh Kishani Sabha (BKS) (both discussed in Chapter 3), and the international farmers' network to which they both belong, La Via Campesina, organised the Climate Caravan, drawing upon one of the primary means of extending your reach used by the PGA network.

The purpose of the Climate Caravan was to educate and mobilise vulnerable peasant communities engaged in land occupations about the interrelated issues of food sovereignty, gender inequality and the effects of climate change, and facilitate networking connections in the form of movement-to-movement communication and sharing of experiences and strategies, and in so doing deepen and extend solidarity networks of grassroots movements in South Asia.

During my years working with PGA Asia, I had developed ongoing work and trust relations with the president of the BKF and some of the key activist cadres who perform important mobilising roles in the processes of land occupation in Bangladesh. While in Bangladesh I have travelled with these cadres to land occupations, and attended organising meetings concerning land occupation with them. I was involved in devising, helping to raise funds for, documenting and participating in the Climate Caravan, speaking at and facilitating many of its workshops and seminars. The Climate Caravan was particularly pertinent to farmers' movements in Bangladesh, where, as I discussed in Chapter 3, the challenges of climate change fold into existing conflicts over access to key socio-environmental resources such as land, from which the poor have been largely excluded.

In Bangladesh, the poor quality of the road infrastructure, infrequent bus transport, and monsoon weather (which can make roads impassable) makes the work of fashioning face-to-face connections between movement actors difficult and time consuming. Activist cadres and local leaders use mobile phones to maintain information flows and intra-movement connections, but such means of communication cannot fully compensate for the critical moments of connection and interactivity that only face-to-face meetings can generate. This was why increasing awareness about issues such as food sovereignty and climate change at the village level was a key aim of the Climate Caravan.

The Climate Caravan embodied a politics of staying mobile, generating place-based encounters and connections between differentially positioned activists. The Caravan comprised three buses containing 80 activists: 55 BKF and BKS activists from various districts from Bangladesh, and 25 activists from various grassroots movements and groups beyond Bangladesh,[28] meeting with BKF/BKS-organised peasant and indigenous communities in 18 villages in 12 districts of northern and southern Bangladesh, involving a total of approximately 3,000 peasant farmers. In this sense, the Climate Caravan itself was a convergence space of activists from different parts of the world, and the events that it generated along its route also comprised spaces of convergence of these activists with land occupation communities organised by the BKF and BKS.

Activist cadres of the BKF and BKS participated with local movement leaders in mobilising land occupation communities to host the Climate Caravan. In so doing they drew upon local leaders' knowledge of place, in terms of the most appropriate and resourced sites for hosting the Climate Caravan. BKS activists' knowledge of where to source the best local food was deployed, as was how best to accommodate Climate Caravan participants, and how to rig up electricity connections for lighting and phone and laptop charging. The Climate Caravan took place in sites of production and social reproduction – moving between peasant land-occupation sites and spaces of livelihood. It utilised compositional and organisational power by making some space for encounters between activists such as face-to-face interactions during meals. It also created sites of potential such as teach-ins and workshops where the possibilities of food sovereignty practices as means of adapting to climate change could be discussed .

Such encounters generated relational power promoting dialogue, mutual learning, trust and the sharing of informational resources between farmer activists from different place-based movements.[29] They also produced connections through which movement ideas could be diffused or transmitted regarding grievances, the goals of social change, organisational development, strategic assessment and so on.[30] Through communication, cooperation and coordination with peasant communities, the BKF and BKS sought to territorialise the movement both within and beyond the immediate spaces of its land occupations. Here, the practice of solidarity was at once specific enough to mobilise and empower at specific territories of occupation, and mobile enough to generate common ground between communities nationally and internationally.[31]

However, the ability of activists to stay mobile is uneven, as previously mentioned. There are pronounced differences in physical mobility across space and access to resources (such as information, education, time, money, technology) between participants. Poor, land-occupying peasants were less resourced than movement activists crossing national borders to participate in the Climate Caravan, national BKF and BKS leaders based in Dhaka, and activist cadres who travel throughout Bangladesh organising village communities (see Chapter 3). Some circulate more freely and extensively than others and are differentially empowered, reflecting class, gender and caste hierarchies within South Asian societies.[32]

Nevertheless, the Climate Caravan acted to deepen the organisational power of the BKF and BKS, informing, consolidating and extending territories of occupation in different ways. First, the connections forged as a result of the Climate Caravan increased the cohesion between BKF and BKS members from different districts in the country. This was facilitated through local leaders and some peasant farmers from different northern territories of occupation joining the Caravan, and in so doing meeting activists from southern territories of occupation and discussing their experiences during the Caravan's events. As one BKF activist commented to me after the Caravan: '[T]he Caravan was able to make a bridge between people in the north and south ... to facilitate greater mutual understanding.'[33]

Second, the participation of activists from farmers' movements from India, Nepal, Sri Lanka, Pakistan and the Philippines provided an opportunity for peasants to share experiences from their different movements' struggles and national contexts. The Caravan also allowed participants to meet with Bangladeshi peasants, explore how they might create bilateral campaigns with the BKF and/or BKS, fashion joint campaigns with other movements and take their experiences back to their own countries and struggles. Such connections enabled the translocal diffusion (of ideas, tactics, strategies and so forth) between different sites and social actors, bridging cultural and geographic divides,[34] and facilitating solidarity between movements, as an Indian activist commented:

We have formed relationships, deepened networking ties, and we have begun to plan future actions together. I think it was encouraging for communities to see an international presence, and that others care about the problems of people in Bangladesh and want to learn from them. This is solidarity.[35]

Further, the connections fashioned through the political activities of the Caravan can have impacts on politics in particular places in productive ways. For example, one local BKS leader in a territory of occupation visited by the Climate Caravan had previously experienced police violence and harassment by local government officials. A Caravan event in her village enabled her to debate with a local government

Figure 6.1 Extend your reach: peasant activists from Bangladesh, India, Nepal and Pakistan lead a Climate Caravan demonstration in Dhaka, Bangladesh, 2011. Photograph by the author.

official from a position of relative empowerment, owing to the presence of both national movement leaders and activists (including a member of parliament) from other countries. The BKS activist informed a BKF national leader (who informed me) that after the Caravan had left the village she was contacted by the official who apologised for the 'problems' she had faced in the past and suggested that they work together in the future.

However, the Caravan was also an example of the entangled powers of resistance and domination discussed in Chapter 1. Gendered responsibilities influenced the level of women's participation in the Climate Caravan – when and where women were able to participate – and over-determined the form of their participation. In most of the Caravan seminars and workshops, men comprised no less than 70 per cent of the participants, although there were significant levels of female participation at the four public rallies. The timings of Caravan meetings were frequently inconvenient for women, owing to the gendered division of labour that positions them as housewives (rather than workers or activists) and requires them to cook for the family as well as look after children and attend to unforeseen events such as sickness, family problems and so on. In addition, as some activists mentioned to me, women's participation was also constrained by everyday social relations such as their lack of decision-making power within community contexts.

Beyond participation in the mass rallies and a few of the workshops, much of BKS women's participation in the Climate Caravan involved working in the kitchens or was compromised by family responsibilities. While sourcing and preparing food is crucial reproductive labour, the gendered division of labour reflects the predominance of gendered organisational models of leadership within South Asian peasant movements that favour charismatic males,[36] something which continues to compromise the level of women's participation in spaces of discussion and education such as workshops and seminars. This was noted by a prominent BKS activist during the Climate Caravan:

> The Kishani Sabha [BKS] contributed a lot to the Climate Caravan, through the participation of local leaders. We purchased food and did the cooking. I was involved in cooking but this was an important part of the Caravan … [T]his is practical food sovereignty. We expected more female participation but we did not get as many [women] as we expected. Some leaders were invited to come with the caravan team but they could not make it; many women leaders from the north were not able to join the caravan team because of family problems.[37]

What these various problems attest to is that the seamless extension of reach implied by digital technologies must always confront the friction that exists in place-based material struggles. Hence *placed* practices of political articulation such as the meetings, workshops, rallies and demonstrations of the Climate Caravan became critical moments of translocal connection between territories of occupation in Bangladesh and social movements from other South Asian countries.[38] Because of the considerable planning, resources and coordination involved, such practices are intermittent. Hence, a follow-up to the Climate Caravan only took place in 2014, taking activists from different farmers' movements through Bangladesh, India and Nepal.

EXTEND YOUR REACH

While social media enables the spreading of activists' wars of words, grounded political actions remain critically important, both in placed struggles (such as that of the BKF) and in attempts by activists to fashion solidarities with others. Global days of action, conferences, caravans and other initiatives that enable face-to-face meetings between activists can enable the sustainability of activist and movement identities, and practically and symbolically articulate the common ground shared by different placed-based social movements. Moreover, in these actions places become 'articulated moments'[39] in the enactment of global networks. As a result of these types of action, there are differential impacts on particular place-based struggles, due in part to the extent to which a particular struggle is projected onto the global arena by virtue of its involvement in a globalising network. However, certain scales of political action may provide more appropriate means for movements within convergence spaces to measure their strength and take stock of their opponents than others. For example, many movements in the Global South see the defence of local spaces and opposition to national governments (pursuing neoliberal policies) as their most appropriate scales of political action.[40]

I have shown how international networks of solidarity have their roots in local struggles in a variety of places, and that in order to act globally one must also think locally. Moreover, struggles that appear local in character are influenced by, and in turn have the potential to influence, conditions, decisions and processes that occur far beyond the local; in other words, to act locally one needs to think globally as well. These different politics of scale provide movements with a range of opportunities and constraints. For example, problematic issues continue to arise within convergence spaces concerning unequal discursive and material power relations that result from the differential control of resources and placing of actors within network flows. These in turn give rise to problems of representation, mobility and cultural difference (especially gender), both between the social movements that participate and between activists within particular movements.

While activities to promote solidarity generate emotional energy and connections between activists, a further challenge confronts them (in some ways exemplified by the gendered inequalities that confronted the Climate Caravan). This concerns the challenging of everyday assumptions of activists and everyday people in wider society and opening up potential for people to think, feel and act differently. It is to this that I now turn.

7

Feel Out Of Place

Ethical Spectacles, Zaps and Guerrilla Performances

'Feel out of place' refers to how particular activist practices challenge everyday assumptions by opening up potentials to think, feel and act differently, and by challenging the meanings and feelings associated with particular places. Such practices engage with critical emotional and thinking responses in activists, the police and the public that can alter the landscape of protest in important ways. They are important for the prosecution of protest because, as noted in Chapter 1, dominant or hegemonic ideas in society are propagated by politicians, corporate advertising and the mainstream media, and frequently become the generally accepted practice and way of life for the majority within society. Because hegemony is enacted though lived practices, transforming hegemonic social relations requires challenging dominant ideologies (what Antonio Gramsci termed 'common sense') as they are experienced in everyday practices.[1] Gramsci realised that transforming consciousness within society (such as challenging people's assumptions about a variety of issues) involved action in the realm of culture, and that one of the primary ways to achieve this was through engaging with people's emotions.

Emotions are a means of initiating political action. People become politically active because they feel something profoundly – such as injustice or ecological destruction. This emotion triggers changes in people that motivate them to engage in political action. It is people's ability to transform their feelings about the world into actions that inspire them to participate in political action.[2] Emotions are both reactive (directed towards outsiders and external events) and reciprocal (concerning people's feelings towards each other). Collective activities and interactions associated with protests can generate feelings of togetherness, personal attachment and emotional intensity.[3] Such shared emotions of activism create shared collective identities and are mobilised strategically to generate motivation, commitment and sustained participation.

Activists create shared emotional templates in order to find common cause and to generate common narratives.[4] Indeed, activists use emotion-laden rituals to engender solidarity and self-transformation among participants to inspire and sustain activism. Distinctive emotional cultures are created that influence how emotions are expressed and managed when protestors engage with the public, political elites and opponents.[5] For example, the National Organisation of Women in the United States used feminist consciousness-raising among women to transform feelings of frustration, hopelessness, alienation and anger into a sense of injustice that helped to promote collective action.[6]

Performance and the performance of emotions have become increasingly important in the practice of politics. Indeed, emotions have always been an important element of the practice and performance of politics through the engineering or channelling of fear, anger, aggression and so on.[7] Within protest events, there are a range of 'feeling rules' whereby potentially disruptive emotions (and the political risks that they embody) are emotionally engineered by the authorities. Certain feelings and behaviour are considered legitimate at protests and others not.[8]

Protestors attempt to undermine 'cementing emotions' that underpin extant relations of domination such as fear, self-doubt, shame and hatred while reappropriating, redirecting and intensifying other emotions such as pride, anger and hope in order to bring about social change and change the feeling rules of everyday situations.[9] For example, sorrow and grief moved hundreds of 'grieving mothers' to break the silence concerning political disappearances, torture and extra-judicial killings by the Argentinian junta during the 1970s. As mentioned in Chapter 1, Las Madres de Playa de Mayo deployed such emotions to shame the regime, and they have subsequently been deployed by grieving mothers around the world. Hence, starting in 1988, Israel's Women in Black deployed their shared grief about war victims and their moral outrage about warmongers to engage in protest and defy the government ban on showing solidarity with Palestinians.

Emotional interventions, then, are frequently aimed at sites of assumption and try to engender 'feeling out of place', whereby activities in particular places disturb and challenge the everyday understandings of those places. For example, during the Civil Rights Movement in the United States, mixed-race groups of people violated the law and sat at lunch counters demanding to be served, challenging assumptions about segregation and in so doing generating sites of potential by prefiguring the world they wanted to live in, one without racial discrimination where all people could sit together without being segregated according to race.[10]

Emotional interventions can be potentially transformative because contemporary media present the real world as a drama or staged spectacle, the manipulation of media images constituting a key process within contemporary politics.[11] Under such circumstances, both politicians and grassroots activists 'act' for television, hoping to elevate their actions into public events.[12] Activists perform various types of theatrical politics deploying dramatic or unexpected imagery to make a deeper impression on the viewer.

To discuss these themes, I examine three examples of activism that challenge the meanings and feelings associated with particular places. First, I discuss the alter-globalisation protests of the Clandestine Insurgent Rebel Clown Army (CIRCA) at the G8 meeting in Gleneagles, Scotland, in 2005, which drew upon and developed the emotional registers of activists and was enacted at particular places of performance. Second, I discuss the politics of identity associated with Lesbian, Gay, Bisexual, Transsexual and Queer (LGBTQ) activism that subvert or open up the emotional registers of particular places. Finally, I discuss performances by the Russian feminist group Pussy Riot that co-opt the use of public space for particular performative interventions.

REBEL CLOWNING AS ETHICAL SPECTACLE[13]

In early July 2005, the G8 met at Gleneagles to hold their annual summit, at which top government officials discuss issues such as the macroeconomic management of the global economy, terrorism, arms control and so on. A classic site of decision, this summit was accompanied by alter-globalisation protests that challenged the 'business-as-usual' performance of such summits by attempting to disrupt their operation (as discussed in Chapter 6).

In the months prior to the G8 protests, a collaboration between artists and activists from the Laboratory of Insurrectionary Imagination (see Chapter 4) toured nine British cities and held workshops to train activists in the tactical performance of CIRCA, which had been founded in November 2003 to respond to a visit to the UK of the then US president, George W. Bush.[14] In early July 2005, people from around the world gathered in Edinburgh for the 'Make Poverty History' march and a week of actions around the G8 meeting in Gleneagles, and were confronted by an army of almost 200 clowns.

Stephen Duncombe has argued that creative, culturally attuned political interventions comprise 'ethical spectacles'.[15] These are tactical performances whereby activists make some space to raise public awareness about issues and intervene in sites of assumption to challenge popular understandings, challenging the 'common sense' espoused by the elite and opening up a dialogue on a particular issue. For Duncombe, ethical spectacles are forms of culture jamming: they are theatrical, participatory (in that they include the public in their performance) and transparent, creating clearly absurd spectacles to get people to reflect upon normal reality in some way. They also involve transformative play (such as humour and satire) to communicate messages to the public, and require staying mobile (being open to ongoing modification and adaptation to specific situations). The idea behind transparent spectacles is to allow spectators to look through what is being presented by the performers to the reality of what is really there. CIRCA represented a form of ethical spectacle that brought together non-violent direct action with practices of clowning to create disruptive and emotive interventions in specific places of protest.

The aim of CIRCA was to develop a methodology that transformed and sustained the inner emotional life of activists as well as being an effective technique for taking direct action. The group argued that there was a destructive tendency within many activist movements to forget the inner work of personal transformation and healing in addition to practising politics on the streets.[16] The explanation behind the group's name and acronym was explained on its website:

> We are clandestine because we refuse the spectacle of celebrity, and without real names, faces or noses, we show that our words, dreams, and desires are more important than our biographies. We are insurgent because we have risen up from nowhere and are everywhere, and because an insurrection of the imagination is irresistible. We are rebels because we love life and happiness more than 'revolution', and because while no revolution is ever complete, rebellions continue forever. We are clowns because inside everyone is a lawless clown trying to escape, and because nothing undermines authority like holding it up to ridicule. We are an army because we live on a planet in permanent war – a war of money against life, of profit against dignity, of progress

against the future. We are an army because a war that gorges itself on death and blood and shits money and toxins, deserves an obscene body of deviant soldiers. We are circa because we are approximate and ambivalent, neither here nor there, but in the most powerful of all places, the place in-between order and chaos.[17]

CIRCA confronted the hegemonic (militarised) political discourse and practice of the war on terror – initiated by President Bush following the 9/11 terrorist attacks in the Unites States – and subverted it through creating an army of clowns. The clown is a popular archetype seen at circuses and children's parties, in hamburger adverts and on public streets. While clowns are frequently objects of fun, entertainment or commoditisation, they are also subversive figures who confuse categories imposed by the system and undermine authority by holding it up to ridicule.[18] As such, clowning represents a challenge to processes of governmentality that attempt to generate normalising behaviours and regulated conduct amongst people.[19]

An important performative element of clowning is that of play. Play can rearrange and question existing social arrangements, and through parody and satire, connect bodies, emotions and lived worlds.[20] Play also has a transformative dimension: satire, irony and humour can make political messages more palatable to the public and encourage active participation from the audience, as well as build relational power.[21]

Activists' embodied performances at protests can be emotionally potent, not least because, under protest conditions of danger and uncertainty, they can introduce the element of play.[22] CIRCA actively made some space through embodying an emotive politics that was deliberately disruptive and challenged the performative logics of the G8's stage-managed political event in Gleneagles. CIRCA disrupted the 'feeling rules' of protests between the rebel clowns themselves, and between them and other protestors, the public and the police. The articulation of particular emotional registers was a specifically *placed* activity: the efficacy of CIRCA's performances depended upon the production of an emotive politics in specific locations.

Multiforms, Manoeuvres and Rebel Clown Logic

CIRCA was underpinned by a 'rebel clown logic' – an associative logic, based on visual signs, wordplay and emotional interventions – that served as both a critique of the dominant discourses of the G8 and a challenge to the policing of protest space. The rebel clown logic of CIRCA was constructed through a range of practices. These included workshops that trained rebel clowns in different forms of collective movement (rebel clown manoeuvres) and a range of games that were developed to enable activists to 'get in clown' – that is, to tap into those senses of spontaneity, complicity and absurdity associated with the subversive intent of rebel clowning.

First, both individual activists and the different rebel clown affinity groups to which they belonged took humorous names that mocked military rank and practice. For example, activists named themselves Private Parts, Corporal Punishment, Major Disaster, General Strike and so forth, and their affinity groups took such names as Glasgow Kiss (the group that I participated in, named after the Glaswegian slang for a headbutt), Group Sex and Backward Intelligence. Second, all CIRCA 'clownbattants' shared a common 'multiform'. Military uniforms were deconstructed, decorated and subverted according to the individual creativity of each person and/or group. Third, rebel clowns wore personalised make-up that hid activists' true identities, providing protection from police surveillance and exaggerating the absurdity of the notion of a clown army. While military uniforms are associative of war/security (in the form of the regular army) and aggression/ militancy (in the form of certain types of grassroots activism), the (deconstructed) rebel clown multiform in combination with the usually friendly clown face acted as forms of tactical media, attempting to undermine the intimidation and violence associated with the policing of alter-globalisation protests.

Rebel clowning utilised the emotive power of transformative play (such as humour, satire, mimicry, wordplay and surprise) to communicate messages of opposition to the G8, subverting protest 'norms' and altering the emotional dynamics of protesting. Rebel clown logic drew explicitly on key elements of clowning, attempting through playful confrontation to exaggerate and invert the social order, recontextualising it in order to reveal its absurdity and inviting others (such as the public) to reconsider it. Rebel clown logic was combined with the multiforms, clown faces and clown manoeuvres in order to intervene in sites of assumption and attempt to subvert the hegemonic logic and taken-for-granted world articulated by the G8.

Rebel clown logic used (deconstructed) army uniforms to associate itself with a culture of permanent war, and through that to the connections between neoliberalism, wars for oil and the war on terror. Through the power of association – under the rubric of the 'war on error' – rebel clown logic, combined with the performance of a clown army, waged a war of words that parodied the war on terror. The 'war on error' was a struggle waged against the political 'errorism' of imperial wars (such as the war in Iraq), the economic 'errorism' of neoliberalism and the environmental 'errorism' of over-consumption and fossil-fuelled economies.

The 'war on error' served as an emotional domain of desire, laughter and joy that attempted to overpower the dominant discourse – that the G8 was meeting to discuss and solve the problems of war, debt, injustice and environmental crisis – through a range of tactical performances. For example, CIRCA's Operation HA.HA.HAA (Helping Authorities House Arrest Half-witted Authoritarian Androids) was deployed to invert the logic and expectations of the global day of action against the G8. The choreography of prior protests against neoliberal globalisation had included some activists utilising the 'black bloc' tactical regime of dressing in black, covering their faces with bandanas and engaging in property damage and violent confrontation with the police, with others trying to climb security fences and disrupt meetings. In contrast, CIRCA wanted to deploy rebel clowns to keep the world's most dangerous 'errorists' (the G8 politicians) under house arrest in perpetuity, by helping the police build the fences even higher around their meeting place at the Gleneagles Hotel, and never letting them out.

Through the idea of a clown army (and its resonance with the wars in Afghanistan and Iraq) and the satirical representation of the war on terror, the public were invited in on a variety of jokes in order to create an emotional intimacy between them and the rebel clowns. CIRCA was clearly not a real army but rather people who were presenting themselves as one, albeit a subverted, deconstructed, deviant clown army. It was an obvious performance, and in particular, one that alienated the familiar world of militarisation and foreign wars. In so doing, it encouraged the viewer of the spectacle to step back and look critically at the taken-for-granted world.

The patent artificiality of the 'war on error' made the message more effective. It caused people to laugh (thereby displacing fear of 'security' at demonstrations); it highlighted the falsity of supposed reality (that the G8 leaders could really solve the problems associated with neoliberalism and war) and it let the audience in on the production by dressing up as an army of clowns.

The main target of protestors is often a potentially mobilisable public rather than decision-makers.[23] CIRCA's strategy was to acknowledge the vulnerability of activists as well as people's fear and concerns during protests and in the climate of tension generated by the war on terror. Part of this strategy was to intervene in a site of potential and attempt to transform fear through laughter, play and ridicule. CIRCA actively sought to undermine and ridicule the intimidation and provocation practised by security forces at demonstrations through a strategy of 'divide and fool'.[24] Rebel clowns blew kisses to riot police behind their shields (in Edinburgh at the Carnival for Mass Enjoyment, one rebel clown was seen landing a big red lipstick kiss on a riot shield). Such action feels out of place in everyday images of protest and destabilises mainstream discourses that juxtapose protestors as angry and violent with policemen as peacekeepers.

Thus CIRCA consciously attempted to disrupt protest feeling rules through a range of practices. These practices gained in emotional power because of their place of performance at the site of decision of the G8. Here, disruptive expressions of emotion were controlled. Moreover, they were enacted in full view of the public and the media, hence amplifying their particular emotive effects. CIRCA were not just occupying space in order to protest but also opening up space to alternative experiences, feelings and meanings, hence politicising it.[25] CIRCA was also claiming space through emotional bonds,[26] with spaces of (police) intimidation transformed into clown spaces of play and mockery. These were facilitated by the ability of CIRCA to stay mobile during the protests, adapting to circumstances as they emerged, intervening at sites of circulation of the protests through rebel clown manoeuvres.

For example, in order to mock stop-and-search 'anti-terrorist' laws, which have regularly been used to intimidate protesters, CIRCA clown-battants filled their pockets with deliberately ridiculous objects such as strings of sausages, feather dusters, underwear, rubber ducks and sex toys. In the event of a stop-and-search by the police, such items would have to be laid out on the street and documented by the police. Further, CIRCA's actions helped to diffuse tensions at certain times and places during the protests, as one activist commented to me after my affinity group intervened at a site of destruction, the Faslane nuclear submarine base:

There had been a real sense of intimidation and threat from the police at the base during the early part of the morning. This all changed when you clowns turned up. The police didn't know how to respond to you. There was that hilarious moment when a group of clowns crept up behind a police chief and started following him around, copying his movements. Every time he turned around they would all freeze until he turned back and continued walking. We were all cracking up laughing. You really changed the atmosphere at the base.[27]

CIRCA also subverted the usual relationships between protestors and police. The policing of protests in the UK involves a co-performance between police and protestors. The police operate with specific 'frames' that influence how they view different protest groups, such as differences between legitimate and illegitimate protest, and 'good' and 'bad' protestors.[28] Indeed, the police stereotype activists into three groups: genuine protestors, 'troublemakers' and 'rent-a-mob'.[29] In addition, emotional control scripts apply as much to police as they do to protesters – even if certain police exhibit excessive behaviour such as beating, or even causing the death of activists. In the broader geopolitical terrain of the war on terror, and under the heightened security that attended the G8 meeting, the police were expected to maintain a particular 'restricted' emotional register, not least to reduce the political risks of disruptive emotions in those responsible for policing the protests.

However, rebel clowns disrupted these emotional scripts. For example, during the demonstration at Gleneagles during the global day of action,

Figure 7.1 Feel out of place: the Clandestine Insurgent Rebel Clown Army, Faslane submarine base, Scotland, July 2005. Photograph by Rabbitman.

protestors were hemmed in and delayed for long periods by the police. There seemed to be a deliberate attempt to antagonise the crowd. My affinity group, Glasgow Kiss, decided to take action to change the dynamic. We noticed that there was a private house set back from the road. The front garden was separated from the street by a waist-high wall. Behind the wall were ten police officers, all on alert, with grim set faces and stiff bodies, bristling with latent aggression. Accompanied by Captain Outrageous, I went over to the wall and started playing peek-a-boo with the police. The rest of Glasgow Kiss joined in. We would duck behind the wall and then jump up in front of the police officers just the other side of the wall and shout 'Boo!' at them. This went on for a while. At first, the police just stood there staring blankly ahead. Slowly, their faces began to crack and emotions that were out of place in the performance of policing at the G8 began to emerge. A smirk here, a grin there, some stifled chuckles. A female officer started to really laugh and had to be replaced by the commanding officer.

However, at other times the police became angry, frustrated and annoyed by the mockery of the clown army activists and acted in an aggressive manner. This is in part because rebel clowning frustrated police management protocols, since they are dependent upon negotiating with protestors (talking sense to them) rather than dealing with a rebel clown who, while sensitive, acts nonsensically.[30] An example of this occurred when a police officer approached a group of rebel clowns and asked them who was in charge. Rebel clown logic necessitated each clown point in a different direction (including up and down) to no one in particular and shout 'he is', 'she is', 'they are', 'the dog is' and so on. Also, as one function of mounted police is to intimidate protesters (and thus generate fear), the act of laughing at the police acted to dispel fear, redistributing power and agency in particular situations and places. It subverted the protocols of policing and deconstructed the opposition between 'good' protestors who obey authority and 'bad' ones who react with violence.

Following the 2005 G8 summit, clown armies appeared in the United States, Europe and Australia amongst other places. However, activists were frequently undertrained in clowning techniques, attracted more by the spectacle of rebel clowning than its subversive content. As a result, the culture jamming potential of rebel clowning has been dissipated. However, the key elements of ethical spectacles remain potentially potent tools in activist practice. Those elements that facilitate feeling out of place continue to appear in creative forms of protest such as LGBTQ activism and Pussy Riot.

LGBTQ ACTIVISM

'Queer' is a term often used to denote all those 'othered' (and rendered out of place) by heterosexual society, while in its more radical forms queer celebrates gender and sexual fluidity (rather than homosexuality), and consciously blurs identity binaries.[31] To be inclusive of such difference, this section discusses protest prosecuted by Lesbian, Gay, Bisexual, Transsexual and Queer (LGBTQ) activists.

The LGBTQ community has a long history of disturbing taken for granted notions of space and what is deemed by mainstream society as appropriate behaviour and feelings associated with particular places. For example, on 28 June 1969, New York City police raided the Stonewall Inn, a gay bar in Greenwich Village. In what became known as the 'Stonewall Riots', gay people, along with drag queens, people of colour and young people responded with angry protests rather than passive acceptance of the law.[32]

During the AIDS crisis of the 1980s, the gay liberation movement in the United States used pride (embodied in Gay Pride marches and celebrations) to submerge anger, remember those who had died from AIDS and encourage activist participation. However, as political opportunities within the mainstream became increasingly constrained, groups such as ACT-UP (AIDS Coalition to Unleash Power) engaged in more confrontational tactics. These more oppositional practices involved open displays of emotion, such as public grieving by participants.[33]

Engaging in culture jamming, ACT-UP reclaimed and used the symbol of a pink triangle, which had been originally used to identify homosexual prisoners in Nazi concentration camps. As part of their war of words, ACT-UP used the slogan 'Silence = Death', placed above the pink triangle to call attention to the severity of the AIDS crisis – often underreported by the media – so that its impact upon the gay community could to be seen and heard.[34]

In 1990, in New York City, the organisation Queer Nation emerged out of the protests and actions of activists from ACT-UP in response to escalations of anti-gay and lesbian violence on the streets and prejudice in the media and arts. One of the key methods of direct action employed was to practice a politics of visibility in 'straight' spaces in order to call attention to the fact that most public space was, in terms of its norms and practices, heterosexual space. For example, a war of words on public banners and stickers (such as 'Cock Sucking Faggot – Queer Nation') sought to reclaim anti-gay language, while straight bars would be frequented by Queer Nation activists who would engage in public displays of affection parodying straight behaviour. The purpose of such actions was to make it clear that queers would not be restricted to gay bars for socialising, and publicly question the naturalised status of heterosexual coupling activities. Such interventions in public spaces transformed those spaces into sites of assumption and potential. Everyday assumptions concerning 'acceptable' displays of affection in particular public spaces were challenged, and in so doing those spaces were opened up to the potential of becoming more liberated and less discriminatory.[35]

Another tactic popularised by the early gay liberation group Gay Activists' Alliance in the United States (and subsequently used by Queer Nation amongst others) has been the zap: a form of direct action consisting of a raucous public demonstration designed to embarrass a public figure or celebrity while at the same time calling attention to particular LGBTQ rights and issues. While a range of other tactics could be used in a zap, such as an occupation of space, street blockades at sites of circulation and the disruption of public events, the main purpose was to create a visible public presence of LGBTQ people in what would otherwise be deemed heterosexual space, and in so doing make straight people feel uneasy and raise particular issues concerning anti-queer discrimination.[36]

A particularly popular zap tactic has been that of the kiss-in, staged in shopping malls and on street pavements as a shock tactic directed at heterosexuals to raise awareness concerning what is deemed 'appropriate' public displays of affection, thereby challenging straight norms or discriminatory practices and laws. For example, in 2014 in Brighton, a 'Big Kiss-In' was held in a branch of Sainsbury's supermarket after a lesbian couple had been threatened with expulsion by a security guard for briefly kissing in one of the aisles. Their affectionate behaviour had been deemed out of place by the security guard and the customer who had complained to her. By filling the supermarket with kissing same-sex couples, the kiss-in transformed an everyday site of consumption and the emotional behaviour associated with it.[37]

Nevertheless, even within the LGBTQ community there were those who felt out of place. The Lesbian Avengers were a direct action group who emerged in New York City in 1992, frustrated at the invisibility of lesbians in mainstream society, as well as their invisibility in the LGBTQ community and their experiences of misogyny within that community. Together with the ACT-UP Women's Network, they created the first Dyke March, which consisted of 20,000 women marching on Washington in 1993 to increase lesbian visibility and activism. Appearing en masse in public, and refusing to apply for a legal permit to march, the women consciously sought to be seen and heard in the US capital.[38]

Queer activism has also emerged to further displace taken-for-granted attitudes within mainstream society as well as dominating attitudes within the LGBTQ community. As gay and lesbian rights activists engaged in sites of collaboration with the state to gain greater respectability and legal status, gender variant people, transvestites and drag queens were seen as an embarrassment and a liability in the cause of assimilation into mainstream society. Queer activism emerged in part as a result of queers feeling increasingly marginalised within the very community that they were supposed to be part of, and has taken three forms.

First, queer activism intervenes in sites of assumption concerning 'controversial' sexual practices such as pornography, prostitution and promiscuity that continue to be considered out of place by a mainstream no longer shocked by same-sex desire. Second, queer activism critiques sites of collaboration that have seen the assimilation of gay and lesbians into the mainstream through legislative reforms such as same-sex marriage and adoption and the associated marginalisation of alternative choices and lifestyles within LGBTQ communities. However, others have argued that such reforms are strategic, opening space for other struggles over the freedom for LGBTQ people to be themselves.

Third, queer activists have extended their reach through alliances with anarchists, anti-capitalists and other subversive movements.[39] For example, the Queeruption Network has established queer convergence spaces, squats and parties, and through their slogan 'Queer mutiny, not consumer unity' they have sought to subvert mainstream Gay Pride events arguing that they have become over-commoditised. For example, at the 2004 Pride Parade in London, radical queers under the slogan of 'Queer Mutiny' made some space for themselves in front of the uniformed members of the Lesbian and Gay Police Association dressing as a camp parody of an anarchist 'black bloc'. Deliberately out of place at the parade, radical queers sought to highlight the corporate mainstreaming of Gay Pride events that transformed them from sites of resistance into sites of consumption.[40]

In the United States, LGBTQ activists are increasingly expanding their organisational power to include struggles against police violence, economic inequalities, immigrant deportation, incarceration and public education cuts. For example, the National LGBTQ Taskforce and Southerners on New Ground both work at the intersection of issues of heterosexism, racism, sexism and economic inequality within US society.[41]

PUSSY RIOT

Outspoken in their support of LGBTQ rights and containing at least one member of a sexual minority, the Russian feminist punk rock protest group Pussy Riot emerged in Moscow in 2011 in protest against discrimination against women in government policy. In particular, Pussy Riot critique cultural nationalism in Russia and have focused on the repression carried out by authoritarian regimes concerning idealised assumptions and behaviour in Russian society concerning sexism, sex and family life. The group is characterised by its wearing brightly coloured dresses and tights and balaclavas, and they have engaged in a range of guerrilla performances in public locations that co-opt and challenge the everyday use and meaning of such spaces, transforming them into sites of assumption. Such performances are then edited into music videos and posted on the internet to extend the reach of the group and its message. Moreover, in subverting the uses of public space, Pussy Riot have attempted to generate a certain amount of institutional outrage over their actions.

For example, in one action in early 2012, eight members of the group simultaneously made some space and waged wars of words by performing the song 'Putin Zassal' (variously translated as 'Putin has Pissed Himself' and 'Putin Chickened Out') on the Lobnoye Mesto in Red Square, Moscow, calling for a popular revolt against the Russian government and an occupation of Red Square. The Lobnoye Mesto is a 13 metre-high stone platform in front of St Basil's Cathedral in Red Square, used historically for the announcements of political edicts from the Tsar and also religious ceremonies. Balaclava-wearing women playing punk music and setting off flares beneath a flag emblazoned with the radical feminist symbol of a clenched fist inside the Venus symbol were clearly out of place in such a public site. Through this performance, Pussy Riot were able to publicise its war of words against the Russian government.[42]

Later in 2012, five members of the group staged a performance of the song 'Punk Prayer – Mother of God, Chase Putin Away' in Moscow's Cathedral of Christ the Saviour. The protest was directed at the Orthodox Church leader's support of Vladimir Putin's presidential election campaign. The song alluded to close ties between the church and the security services, criticised the subservience of many Russians to the church and attacked the church's traditionalist views on women. Their performance challenged taken-for-granted assumptions about the role of the Orthodox Church in Russian society and the uses that church space could be put to.

Subsequently, two of the group members, Nadezhda Tolokonnikova and Maria Alyokhina, were arrested, charged with hooliganism and sentenced to two years imprisonment.[43] However, a key tactic of Pussy Riot is the use of 'illegal' or unsanctioned interventions that disturb the everyday meanings associated with particular places in order to create media attention and challenge everyday assumptions. As international support for the imprisoned women spread, music artists such as Madonna and Björk expressed their solidarity by inviting Pussy Riot to perform with them. However, the group articulated their opposition to the capitalist model of art as commodity production, challenging the assumption that music is a site of consumption and stating: '[T]he only performances we'll participate in are illegal ones. We refuse to perform as part of the capitalist system'.[44]

FEEL OUT OF PLACE

This chapter has focused on acts to make people feel out of place – acts that attempt to challenge everyday assumptions, opening up potentials to think, feel and act differently, and which challenge the meanings and feelings associated with particular places. Such practices engage with critical emotional and thinking responses in activists, the police and the public in order to attempt to alter the landscape of protest. In so doing, dominant ideas (or common sense) in society as experienced in everyday experiences or the use of public spaces are challenged and disturbed. Even if only temporarily, transforming consciousness within society (for example, by challenging people's assumptions about a variety of issues) involves action in the realm of culture, and one of the primary ways to achieve this is through engaging with people's emotions, whether that be through humour (as in the case of CIRCA), unease (as in the case of Queer Nation) or institutional outrage (in the case of Pussy Riot).

What is important about ethical spectacles is the element of play and humour, because play necessitates active audience participation and imagination that creates an intimacy between the performer and the audience. In so doing, such narrative interdependency works against social relations of hierarchy and separation (between the 'performer' and the 'audience', for example) and can open up the public to the political content of activists' messages. As Rist notes, 'messages that are conveyed emotionally and sensually can break up more prejudices and habitual behaviour patterns … than intellectual treatises'.[45] Attention to the emotional dimensions of activist experience and (inter)action remains a potentially compelling intervention in the repertoire of political performance, particularly when combined with the confrontational approach of direct action and the subversive power of humour or the mobilising force of anger. It opens up potentials to deregulate conduct and fashion liberatory feeling rules of political action. The extent to which such actions will be important over the coming decades forms part of the conclusion to this book.

8

Space Invaders

Power, Politics and Protest

At midday on 12 December 2015, a swarm of 15,000 red-clad protestors filled the streets of Paris during the twenty-first meeting of the signatories to the United Nations Framework Convention on Climate Change (UNFCCC), otherwise known as COP 21. The Red Lines protest – held at this site of decision – symbolised the minimum policy requirements for a liveable planet, such as meaningful, verifiable and binding greenhouse-gas emission reductions and equitable climate finance for poorer countries. The protestors' counter-discourse – climate justice – to official pronouncements at COP21 was able to be seen and heard despite the declaration of a state of emergency by the French government, banning protests following terrorist attacks in the French capital a month earlier. The protest was an example of a rising wave of social mobilisation around the world that, commencing in the 1990s with the anti-roads and alter-globalisation protests, has gained new momentum since 2006.

Moreover, the wave of social mobilisation has witnessed some of the largest protests in world history: crowd estimates indicate that each of 37 protest events involved 1 million or more protesters.[1] For example, in India during a general strike in 2013, an estimated 100 million people protested against low living standards, inequality, attacks on wages and the need for better labour conditions.[2] This wave of mobilisation responds to and confronts what the Zapatistas term 'the storm': the convergence of the global economic crisis, increased global economic inequality and the associated increase in power of the international financial sector, the global environmental crisis (including climate change), the loss of legitimacy and faith in traditional institutions such as government, political parties, the police and the media, and the transformation of everything into commodities.[3] While articulating resistance to processes of accumulation by dispossession and the colonisation of people's life-worlds, these mobilisations also articulate alternatives to capital and the state.

Grievances have included: economic (in)justice, inequality and austerity policies (such as the Occupy mobilisations and urban commons initiatives in Greece; see Chapter 3); the failure of political representation and political systems given the increasing corporate influence on governments and their privatisation policies (such as the mobilisations of UK Uncut; see Chapter 4); People's Global Action (see Chapter 6) and CIRCA (see Chapter 7); destructive development (such as protests against the M77; see Chapter 3) and the anti-dam struggle of the Narmada Bachao Andolan (see Chapter 5); the failure of governments to respond meaningfully to climate change (for example, the protests of the Bangladesh Krishok Federation and the Bangladesh Kishani Sabha; see Chapters 3 and 6) and Climate Games (see Chapter 4); and indigenous and people's rights (for example, Black Lives Matter and Idle No More mobilisations; see Chapter 4), the Zapatistas (see Chapter 5) and LGBTQ activism (see Chapter 7). The diversity of protestors has included direct action activists, trade unions, women's and farmers' movements, youth and indigenous peoples' groups and members of the middle classes. National governments continue to be one of the primary foci of protests.[4]

Some tangible achievements have been secured through some of these protests, including changes of government (the revolution in Tunisia in 2010, for example), the adoption of new constitutions (such as in Iceland and Morocco), changes to law or policy (for example, the French government repealed a regressive law on new work contracts for youths in 2006), the exposure of government or corporate secrets (such as WikiLeaks) and symbolic achievements, like changes in public discourse (for example, Occupy made income inequality an issue in both national and international debates).[5]

In the face of ongoing crises of finance, climate, food security and political legitimacy, exacerbated by the resurgence of far-right ideologies, there is an increasing importance for resistance in various forms to what the Zapatistas term the 'capitalist hydra', a system that, like the mythological animal, has many heads and is able to adapt, mutate and regenerate itself across geographic space.[6] Despite their diversity, protests pose political interventions in the political order that determines what is visible and what can be said and heard in political discourse and the spaces in which this occurs. Protests are examples of 'dissensus': enactments of alternatives to the current political order of things.[7] They call into question the structuring principles of that order by making visible the inequalities and lack of freedoms (or wrongs) inherent to it.

They are expressions of counter-hegemony: the withdrawal of popular consent to be ruled and the assertion of popular power including its relational, compositional, organisational and at times representational components.

In this book I have drawn upon my own scholar activism research within geography, social movement studies and the broader social sciences, and a variety of case studies from around the world to argue for a distinctive approach to understanding and prosecuting protest. I have shown that protest activities invariably involve a range of spatial strategies and sites of intervention, the precise combination depending upon the specific goals and tactics of a particular struggle and the geographical context in which it takes place. In this concluding chapter, I suggest possible spatial strategic interventions for the forthcoming decades, a period that will, I believe, continue to experience economic, political and ecological crises generated by capitalism.

KNOW YOUR PLACE

Chapter 2 showed how the material conditions of places of work, livelihood and home can generate protests and influence the character of those protests, as well as how activists frequently use the physical and built landscape to shape their struggles. The immediacy and intimacy of place is critical in enabling activists to draw upon and develop their relational power, since placed, face-to-face meetings nurture and ground collaborations across difference.

'Knowing your place' can also enable activists to culture jam key symbolic sites in order to challenge everyday assumptions associated with them. Just as protestors utilise the physical topography of public space (such as streets, squares and rooftops), so knowledge of place can also ground protests in cultural, economic and political milieux that 'speak' to wider audiences. This can facilitate social mobilisation and inform and empower that which is seen and heard in protests. The idea of 'knowing your place' resources activists so that they are able to make informed choices about their terrain of struggle (and hence which sites of intervention to target), the materials they use in their protests and their protest collaborators.

The ability to 'know your place' enables protestors to deploy place-specific knowledge and cultural codes, and powerful culturally recognisable symbols in order to engage public opinion and participation – whether resisting a missile base, waging revolution or resisting colonisation from the home. However, while political action is frequently informed by local conditions, it is also the product of wider sets of relations, processes and connections. For example, climate justice struggles are responding to resource inequalities and conflicts and their exacerbation by the challenges of climate change.[8] Multiple *placed* 'frontline struggles', what Naomi Klein has termed 'Global Blockadia',[9] are being waged across the planet against fossil fuel extraction (for example, oil tar sands, gas fracking and new coal mining projects) and associated infrastructures (such as airports, motorways, pipelines and corporate headquarters) that are the product of processes far beyond the sites of destruction. In this sense, however localised a protest might appear to be, it always gestures beyond itself.[10]

For example, in 2016 the Standing Rock Sioux Tribe successfully protested against a site of circulation of energy flows – the Bakken (Dakota Access) pipeline in North Dakota – that was planned to transport 470,000 barrels of oil each day across North Dakota, South Dakota, Iowa and Illinois. The pipeline was considered to be an environmental and cultural threat to the land and water of the Standing Rock

Figure 8.1   Site of decision, circulation and potential: climate justice 'people's assembly' outside the UNFCCC meeting, Copenhagen, Denmark, 2009. Photograph by the author.

Sioux because an oil spill would permanently contaminate the water supply, and that construction of the pipeline would destroy lands that contain archaeological sites and ancestral burial grounds.[11] As Eryn Wise, media coordinator at the Standing Rock camp, argues:

> As indigenous people we recognise the need not only to preserve the land but also the water, because it's ultimately the life-giver of the entire world. That's why we call ourselves protectors, not protestors.[12]

The struggle also represented an example of the ongoing indigenous resistance to colonial violence that has seen the violation of treaties between the Lakota people and the US government, limiting their sovereignty to federally managed reservations.

While the resistance at Standing Rock was informed by local conditions and indigenous culture, it was also the product of wider sets of relations, processes and connections. For example, the pipeline is being constructed by a subsidiary of Energy Transfer Partners of Dallas, Texas. Because it crosses four states, the state regulators of each state have had to grant permits for the pipeline to be constructed. Further, representatives from 300 of the 566 federally recognised indigenous peoples in the United States, as well as non-native activists including US veterans and people from the Amazon, Peru, Australia, New Zealand and Japan, have travelled to Standing Rock to show their solidarity with the resistance.[13] In addition, tactics that have travelled from beyond the local, such as 'lock ons' (see Chapter 3) and blockades by caravans of vehicles (see Chapter 6) were also used during the protests.

Therefore, it remains important not to make a fetish out of place or its occupation, as I discussed concerning the Occupy mobilisations (Chapter 3).[14] While being placed in communities of struggle is critical, activists need to ensure that the use of space makes both tactical and strategic sense.

MAKE SOME SPACE

Chapter 3 showed how protestors actively shaped places by physically transforming the character of or meanings associated with them, thereby conveying potent symbolic messages and images to society. Cultures of resistance generate productive relations and practices that frequently contain prefigurative political processes. These may constitute the everyday life practices associated with prosecuting protest (such as constructing encampments to resist road-building or dam-building); they may be structural, born out of the need to fashion socially reproductive alternatives as a response to the deprivations of austerity and inequality (for example, constructing urban commons); or they may be both (such as land occupations concerning food sovereignty).

In making space for protest, activists frequently generate simultaneous sites of intervention. By occupying land or recuperating factories, protestors are intervening in sites of production reflecting their struggles to maintain or create sovereignty over the means of livelihood. By attending to the activities, responsibilities and relationships that are directly involved in maintaining protests, protestors are intervening in sites of social reproduction that enable and resource social movements, the protests in which they are engaged and the production of activist spaces. These provide physical and emotional sustenance for activists, and operational infrastructures for social movements. By attempting to actualise alternatives about how to live 'on the ground', protestors are intervening in sites of potential that can stimulate the popular imagination concerning possible future scenarios beyond the capitalist present. In so doing, activists are also intervening in sites of assumption by attempting to change how people think and feel about particular issues that frequently necessitate challenging underlying beliefs and mythologies.

'Making some space' consists of a series of spatially distributed acts and processes: those physically present are always part of more spatially extensive virtual and digital networks of support and organisation.[15] 'Making some space' also includes the infrastructural conditions of staging – the technological means of capturing and conveying an event, as well as visual and audio communication.[16] Acts of protest (including demonstrations, occupations and assemblies) literally make space: they reconfigure the materiality of public space and can produce the public character of that material environment.[17]

For example, in France, a *zone à defendre* (ZAD) at Notre-Dame-des-Landes in the Loire-Atlantique department has been constructed on nearly 5,000 acres of wetlands, farmland and hamlets. This action is part of a protest against a site of destruction and circulation – the construction of the airport of Notre-Dame-des-Landes, planned by French government in partnership with VINCI, the world's largest multinational construction firm. As an integral part of their resistance, this ZAD has practised creative alternatives to resource extraction through establishing sites of production and social reproduction, cultivating wheat and buckwheat and establishing a textile workshop, microbrewery and bakery.[18]

In addition to particular sites of resistance such as the ZAD, creating convergence spaces such as conferences, caravans, protest camps, social forums and popular assemblies will continue to be a critical component of protests because they enable the development and deepening of relational, compositional and organisational powers amongst activists. While 'making some space' can shape and inform the materiality of activist ecology, activists also need to consider the interplay of fixity and fluidity in their protests.

STAY MOBILE

Chapter 4 showed how the interplay of movement and fixity (or deterritorialisation and territorialisation) can enable particular places to be claimed, defended, strategically used and/or abandoned depending upon the strategies and goals of a particular protest. 'Staying mobile' also refers to the necessity that activists continually adapt to changing contexts and conditions, and can enable protesters to evade capture, deploy compositional power in order to make space and keep strategically ahead of their opponents in particular struggles.

The mobility of the swarm lends itself in particular to interventions in sites of circulation (such as roads, motorways, airports and ports) and can enable potent messages of resistance to be communicated to governments, corporations and other institutions. The occupation of public spaces by significant numbers of people performs popular power in visibly striking ways. However, strategies of mobility mean not always being too obviously 'in place'. The dispersed, decentralised activities of the pack, or affinity groups, provide some protection from infiltration by opponents, enabling activists to conduct strategy planning out of the eye of public meetings and surveillance technologies.

Staying mobile also means knowing one's opponents and adapting strategy to that knowledge with the intention of weakening their alliances. The importance of maintaining initiative and adapting to changing political and protest dynamics and conditions is critical. This is because the 'viral' character of protest ideas can often amount to the copying of tactics rather than adapting to particular circumstances:

Repeating tactics, reifying them as the route to liberation, not only creates a vanguard whose actions are supposed to bring us all emancipation, but enable dominant groups to contain and discipline revolt. The effectiveness of the method is diminished, making the act more-or-less symbolic. To counter this recuperation activists are aware of the need to construct new methods and new alliances to remain 'one step ahead'.[19]

An example of such adaptation to changing contexts occurred recently in Spain, where the No Somos Delito ('We Are Not a Crime') movement protested against the Citizen Safety Law of 10 April 2015 (also known as the 'Gag Law', *Ley Mordaza*), which brings strong fines against unauthorised protests or documenting and publishing images of the police. Instead of organising a mass public demonstration, No Somos Delito projected holographic images representing a mass protest swarm in front of a site of decision, the Spanish parliament.[20] In order to be attentive to the dangers of repetition and reification, activists must continually adjust and update their strategies and tactics to keep one step ahead of their opponents and pay particular attention to their discourses of dissent.

WAGE WARS OF WORDS

Chapter 5 showed how activists intervene in and create their own media spaces in order to make demands and create alternative ways of thinking about particular issues. Language is a key terrain of struggle. The common-sense sites of assumption of mainstream political culture can be challenged through words, images, stories, figures of speech and social media, as well as disobedient bodies on the streets. For example, one of the direct action slogans central to the Climate Games protests at the Paris climate change meeting in 2015 (see Chapter 5) was 'We are not fighting for nature. We are nature defending itself'. While highlighting how water, land and the atmosphere are facing destructive incursions by neoliberal capitalism, activist. framings articulated not only anti-corporate and anti-fossil fuel sentiments but also questioned assumptions concerning the separation of the social, environmental and political in everyday life.

'Wars of words' enable protestors to be seen and heard. This is of particular importance in the current conjuncture of 'post-truth' politics, as Summer Brennan argues referring to the election of Donald Trump to the US presidency in 2016:

> If you can colonise the minds of a population with untruths and confusion, you forcibly re-write reality. This is done with stories. It's done with language. How we speak about the world is a reflection of how we see it.[21]

Activist representations of events are important for not only making demands but also creating alternative ways of thinking about particular issues. The use of words and images is an important strategy in the conflict over representations of events between activists, governments, private corporations and the public. In addition, the creative utilisation of activist media can create protest cultures, convey emotions moving people to act and disseminate movement actions; it can also convey ideas and demands, share information, educate and generate solidarity and create memes. As such, 'wars of words' can be a critical tool in generating sites of potential.

However, while particular wars of words can form a common discourse around which protestors can organise, activists need to be attentive to fashioning common ground that does not efface class, gender, race and disability differences. For example, the notion of climate justice, while a common mobilising frame for activists around the world, can sometimes elide the significant axes of difference that exist concerning the uneven distribution of climate change vulnerability, the uneven ability to meaningfully influence climate futures owing to persistent axes of social difference, and the inequities concerning the price of activism (painful, risky, difficult, costly) for those who are not able-bodied, white, middle-class males.[22]

Therefore, 'waging wars of words' must involve striving to generate shared goals and a sense of shared responsibility without effacing difference. One of the ways to do this is through the power of storytelling, which enables activists to learn about themselves and teach one another how to organise for social change. For example, Black Lives Matter and LGBTQ activism have consciously sought to generate common struggles amongst diverse groups of people that incorporate differences of race and ethnicity, sexual identity, legal status and class. Activists must recognise that there is no such thing as ideological purity – a mythical realm of perfect politics and perfect language. Rather, activism needs to embrace constant experimentation, learning and constructive critique.[23] This is because a struggle creates the language in which new relations express themselves. Through their discourses of dissent, protestors must attempt to construct the kind of affinities that can generate common ground across difference and across geographic space.

EXTEND YOUR REACH

Chapter 6 showed how, through the use of physical and virtual infrastructures and communicative technologies, activists craft and sustain networks of solidarity, which involves thinking and acting across geographic space. While social media enables activists to spread 'wars of words' across geographic space more than ever before, grounded political action remains critically important, both in placed struggles and in attempts by activists to fashion solidarities with others.

This is because local struggles are influenced by, and in turn have the potential to influence, conditions, decisions and processes that occur far beyond the local. For example, the Zapatista struggle emerged as a result of the ongoing exploitation of indigenous people in Chiapas, Mexico, and as a response to the free trade agreement signed by Mexico, Canada and the United States. While based in the Lacandon jungle in Chiapas, the Zapatistas influenced solidarity-building throughout Mexico, and had impacts on the country's economy, not least by discouraging international investment in Mexico. Further, the Zapatistas' critiques of neoliberalism and calls for international networks of solidarity against it inspired a generation of alter-globalisation activists, giving rise to networks such as People's Global Action and the World Social Forum.

It is through networked initiatives that facilitate face-to-face meetings and activities between activists – such as global days of action, conferences and caravans – that the hard work of fashioning common ground and trust between different placed-based social movements is achieved. This includes, as noted earlier, recognition of and engagement with, questions of (class, gender, ethnic) difference. Some encouraging examples of this include certain front-line climate justice struggles that have seen 'unlikely alliances' fashioned.[24] For example, in the United States, Americans Against Fracking has developed a coalition of interest between anti-fracking groups, health workers, environmental and faith groups, peace and justice activists and indigenous organisations.[25] Meanwhile, US-based Indigenous Environment Network is fashioning solidarities with Grassroots Global Justice (an alliance of US working poor and African American communities) and the Climate Justice Alliance (which includes indigenous people, African Americans, Latinos, Pacific islanders and working-class whites in the United States).[26]

Political activities of alliance-building necessarily involve a 'process of contagion',[27] the real-time diffusion and amplification of protest activities around the world: a spread of disruptions at sites of intervention that open up the political order to challenge through visible, embodied artic-ulations of resistance. Alliance-building must also invent ways of saying, seeing and being that nurture a politics of affinity, and that engender new forms of collective and inclusive enunciation, engendering new subjects. This will draw upon relational, compositional and organisational powers that imbue spatial strategies with their force of creativity and antagonism. It will also necessitate being attentive to activists' feelings since the spatial extension of protest is brought about through emotional resonance prior, during and after political interventions.[28]

FEEL OUT OF PLACE

Chapter 7 showed how particular activist practices challenged everyday assumptions by opening up potentials to think, feel and act differently, and by challenging the meanings and feelings associated with particular places. They are important for the prosecution of protest, because, as noted in Chapter 1, challenging and transforming hegemonic social relations requires challenging dominant ideologies as they are experienced in everyday practices. Transforming consciousness within society (challenging people's assumptions about a variety of issues, for example) involves action in the realm of culture, and one of the primary ways to achieve this is through engaging with people's emotions.

Attention to the emotional dimensions of activist experience and (inter)action remains a potentially compelling intervention in the repertoire of political performance, particularly when combined with the confrontational approach of direct action and the subversive power of humour or the motivating force of anger. Indeed, 'mobilising emotions' is a tactic that has been deployed in many of the case studies in this book, such as unease (Queer Nation), shock (Culture Jamming), grief (ACT-UP), anger (Idle No More), grievance (Bangladesh *Krishok* Federation) or institutional outrage (Pussy Riot). Humour, as used by the Clandestine Insurgent Rebel Clown Army, can also be used to undermine or disrupt authority: it can be used to challenge controlling myths, hegemonic ideas and framings in society, enable the construction of group commitment and solidarity (so helping to develop relational and compositional power) and enable people to cope with uncertainty.

By engaging with people's emotions, and by challenging the feelings associated with particular places, activists can target sites of assumption. For example, activist groups such as Liberate Tate, who have protested against the funding of the arts by the fossil-fuel industry, have targeted sites of consumption such as Tate Modern and the Louvre. At the former they covered activists' bodies in molasses, and at the latter they spilt molasses on the floor of the lobby, both acts being attempts to symbolise oil and open up thinking about what Mel Evans terms 'artwash': the legitimation of the oil industry through its broader sponsorship of the arts.[29] In so doing they have challenged what was usually seen, heard and felt in an art gallery. Further, the oil corporation BP recently announced that it is to end its 26 year sponsorship of the Tate.[30]

The targeting of sites of assumption enable what the Zapatistas term 'peripheral vision': an indirect view of an aspect of reality that opens up people's regularised perceptions of places and events to different under-standings. In addition, to construct a more just society, protestors also need to pay closer attention to their (and others') emotions, as well as activists' vulnerability. Activists need to address their own fears and the culture of fear that is being promoted by authorities as a form of social control (for example, concerning terrorism).

Fashioning common struggle will require transforming emotions such as fear and anger into new forms of resistance. According to Subcom-mandante Galeano of the Zapatistas, resistance can deepen and expand through the 'the struggle to transform pain into rage, rage into rebellion, and rebellion into tomorrow'.[31] New activist subjectivities can be fashioned that develop a politics of trust, care and emotional intelligence that nurtures cooperation, social reproduction and empowerment.[32] However, the issue of representational power in sites of collaboration will also need to be addressed.

CHANGING THE WORLD BY ENGAGING WITH POWER

Writing about the radical Left in Latin America, Emir Sader argues that there has been a serious failure to address questions of political strategy and this characterises much of anti-capitalist and some climate justice politics.³³ The autonomist politics favoured by some groups and organ-isations has not enabled a reorganisation of mass forces in order to fashion new liberatory political action and construct alternative forms of political power. Instead, he argues, such autonomist projects have refused to confront the question of power at all – other than stating in rather wishful terms that the world can be changed without taking power.³⁴

It is certainly tempting to consign the role of the state to a form of politics and economics that is only pro-corporate and anti-democratic. Indeed, for many indigenous people the state remains a colonising force threatening their territories with appropriation or assimilation. The post-9/11 crackdown on civil liberties has been accompanied by increased surveillance, the detention of migrants and the rise of increasingly centralised and authoritarian forms of state control and action. Indeed, attempts to marginalise dissent have been accompanied by increasingly punitive policing designed to either dissuade protest or render it ineffectual.

For example, a key element of policing is that of 'invisibility'. First, there is an attempt to render protestors invisible to the public and media, for example through the creation of 'free speech zones' – uncontrover-sial spaces that have little or no visibility to the media or the broader social field, thus rendering protestors' voices silent.³⁵ Second, there is the deployment of 'non-lethal' technologies upon protestors by the police such as Tasers, pepper spray and sticky shockers. These are difficult to capture on the media and leave few visible marks on protestors' bodies for later documentation. Together these processes represent a militarisa-tion of both the police and public space, and an erosion of democratic practices under the guise of protecting democracy.³⁶

Therefore, an engagement with the state only makes sense when progressive social transformation can take place. In this sense, Gramsci's notion of a 'war of position' – that recognises the inseparability of the social, economic and political spheres – remains pertinent, rethinking relations between a radicalised civil society, the economy and the state in an ongoing struggle for hegemony in order to bring about social and ecological transformation.³⁷ This will need to happen across geographic scales and will involve confrontation and antagonism, autonomous projects that develop spaces beyond the reaches of capital relations, and an engagement through sites of collaboration with the state in order to transform it and to upscale local projects.

However, as I noted in Chapter 1, any engagement with institutions must always ensure that challenges to the hegemonic order are not appropriated by the existing system in a way that neutralises their subversive potential. There are some promising examples of where such engagements have born fruit. First, grassroots mobilisations in Latin America and Western Europe have targeted the privatisation of energy and water resources as key sites of struggle and called for their re-municipalisation. Both Denmark's and Germany's renewable energy sectors have seen much decentralised control by local cooperatives.[38] It has also been estimated that between 2010 and 2015, 235 cases of water re-municipalisation have taken place in 37 countries, affecting over 100 million people, especially in the Unites States and France.[39]

Second, in Peru, the indigenous Wampis people have, after collaboration with the national government, established an autonomous indigenous government controlling a 1.3 million hectare territory, bringing together one hundred Wampis communities (10,613 people). In Paraguay, the indigenous Enxet Sur community has, after 18 years of struggle, won legal title from the state to 10,030 hectares of their ancestral home in the Chaco region.[40]

Third, grassroots struggles for food sovereignty by peasant farmer movements – involving calls for peasant control over territory, biodiversity (commons) and means of (food) production – have born fruit in 15 countries, where governments have adopted legislation informed by ideas about food sovereignty.[41] In Venezuela, Mali, Senegal, Nepal, Ecuador, Bolivia and Nicaragua, food sovereignty has been incorporated into national constitutions in some way.[42] In Cuba, the state has supported food sovereignty initiatives,[43] while in the Dominican Republic, Peru, Argentina, Guatemala, Brazil, El Salvador and Indonesia some legislation guided by ideas about food sovereignty has occurred.[44]

However, policy adoption also creates new arenas of social action, since the recognition of grassroots demands by the state does not necessarily imply their full implementation. Indeed, state involvement (for example, in energy re-municipalisation, indigenous rights and food sovereignty) is a dynamic, ongoing and open-ended process of contestation and negotiation, and policy implementation involves a diverse set of state and societal actors in interactive relations that can promote or constrain such alternatives 'on the ground'.[45] Therefore, ongoing mobilisation from social movements remains crucial to the adoption of their demands by the state, both prior to and after that adoption.

Moreover, the case studies in this book also confirm that social transformation cannot and should not wait upon the state to act. Indeed, across the planet there are growing ranks of the excluded, the unemployed, the displaced, the low waged and people in precarious living and working situations (such as migrants). In many countries, austerity programmes mean that the state refuses to meet even basic demands of social reproduction.[46] This is one face of an ongoing crisis of growing economic inequality and legitimacy of political representation engendered by neoliberalism. As a result, grassroots struggles from below remain a critical force that, through the deployment of relational, compositional and organisational powers, create alternatives, and through confrontational collective action can bring the state into a space of engagement and negotiation, rather than domination.

A recent example of this took place in 2014, when Focus E15 Mothers (a group of 29 young single mothers) occupied sites of social reproduction, vacant flats on the Carpenters Estate in Newham, east London, in protest at the local council's plans to evict and relocate them. In a London borough where over 24,000 households are waiting for somewhere to live, the housing estate – containing 600 homes – was earmarked for sale to private investors, who planned to demolish and replace it with high-rent accommodation. The women had been evicted from a local mother-and-baby hostel a year earlier after its funding was cut as part of the council's austerity policies. Council officers advised the women to leave their families and friends in London and move with their children 200 miles away to low-rent areas of Birmingham and Manchester. In addition to the occupation of the flats, activists occupied sites of decision such as the council's offices, and attracted media attention by holding an impromptu party in a site of consumption, the luxury show flat of the housing association that was evicting them. After two weeks of occupation, the council agreed to rehouse the mothers in London.[47]

PRACTISING RADICAL GEOGRAPHIES OF PROTEST

Many of the struggles discussed in this book – such as the land occupations of the Bangladesh Krishok Federation (BKF) and Bangladesh Kishani Sabha (BKS), the anti-dam struggles of the Narmada Bachao Andolan and a variety of other movements like Occupy and alter-globalisation groups such as the PGA – embody struggles for the commons: collective spaces with social relations and norms based on reciprocity, trust and care (rather than individualism, competition and self-interest). Others, such as Idle No More and the Zapatistas, embody struggles for indigenous sovereignty and autonomy. The spatial politics of commons projects contest capitalist enclosures of resources in order to reclaim the public realm for alternative and creative social, economic and environmental practices.[48] As we have seen in this book, these are not just forms of local experimentation – drawing upon relational, compositional and organisational power in a range of different ways – but are also translocal in character, linking up with other struggles in broader sets of (networked) anti-capitalist practices and imaginaries.

Contemporary struggles are informed by, and shape, the geographical and cultural contexts in which they take place through at least five processes: *creative practices* that refashion how people relate to one another and what they do in response to different challenges; *embodied technologies* that merge placed material politics and social media to create more participatory, interactive cultural environments and generate increasing dialogue across geographic space; *transversal collaborations* between diverse placed constituencies that include indigenous people, environmentalists, trade unions and peasant farmers that open up new spaces of solidarity across different geographic scales; *alternative imaginaries* developed through art, digital media and so forth concerning future political, ecological, economic and cultural scenarios and informing attempts to actualise political, economic, cultural and environmental justice 'on the ground'; and a *politics of social reproduction* that stresses the centrality of the activities, responsibilities and relationships (such as caring, affection, communication) that are directly involved in maintaining protests. Indeed, social movement organising frequently involves practices of everyday life as central to the fashioning of alternatives.[49]

For example, the BKF and BKS constitute a national movement that links together a host of locally based occupations and creative practices in agro-ecology, food sovereignty and attempts at non-commoditized social relations. As members of La Via Campesina, the BKF and BKS engage in transversal collaborations with other nationally based farmers' movements across Asia, Africa and Latin America, such as the Movimento dos Trabalhadores Rurais Sem Terra (MST), which has occupied and farmed land in Brazil since the early 1980s.[50]

The idea of the commons embodied in such creative practices represents an alternative global imaginary that builds upon earlier alter-globalisation struggles (as discussed in Chapter 6) and potentially links farmers' struggles over communal land rights, land grabs and food sovereignty with indigenous peoples' struggles for sovereignty and autonomy, and mobilisations such as Occupy and Right to the City.[51] All are concerned in some way with reclaiming and transforming space, time and labour from capitalist processes. Within this global imaginary persist local, place-based knowledge, traditions, collective action and forms of solidarity that remain anti-capitalist and are driven by climate justice concerns. Placed initiatives link up with others in translocal coalitions through embodied technologies that find material expression in placed sites of convergence such as the COP21 protests.

These struggles defend productive work and social reproduction from capital accumulation processes. They threaten the rule of capital because they embody living and working alternatives beyond its disciplinary and exploitative logic. Indeed, a politics of social reproduction concerning the welfare state, care work, education and health care is a crucial site of future struggle. As noted in Chapter 3, everyday forms of activism have emerged, not least as the energy generated by some of the 'squares' movements and popular assemblies has relocated to neighbourhoods and communities to focus on collective practices and institutions for the provision of basic needs of everyday life. For example, workplaces are undergoing 'recuperation' across Latin America, inspired by the MST slogan 'occupy, resist, produce', part of its 'war of words'. In Argentina, 300 worker-recuperated workplaces have emerged as sites of production and social reproduction, including metal and ceramics factories, printing shops, grocery stores, media centres, clinics, newspapers, schools, hotels and bakeries.[52]

However, the politics of social reproduction must confront ongoing gender oppression and inequality within movements. Within capitalist society, Silvia Federici has argued that the gendered division of labour continues to serve as one of the primary means by which capitalist social relations of exploitation are maintained: 'sexual hierarchies ... are always at the service of a project of domination that can sustain itself only by dividing, on a continuously renewed basis, those it intends to rule'.[53] Women are the 'primary subjects of reproductive work',[54] being food providers and producers, and guardians of health and caregiving. Yet gender relations are determined in a social system of production that does not recognise the reproduction of labour power as a source of capital accumulation.

The internalisation and appropriation of the dominant values of sexist capitalist society continues to influence social movements and the communities and workplaces from which they emerge. Indeed, gendered inequalities – and hence everyday struggles against gender oppression – underlie struggles for land and climate justice (such as that of the BKF and BKS), attempts to reclaim indigenous sovereignty and autonomy (such as the Zapatistas and Idle No More), women's rights (such as some LGBTQ activism and Pussy Riot) as well as anti-capitalist struggles (such as Occupy). Activities associated with social reproduction and the oppression of women continue to be crucial grounds of struggle for women as well as men within and beyond social movements. Therefore it is critical for movements to recognise that the 'material means of reproduction is the primary mechanism by which collective interest and mutual bonds are created' and 'the first line of resistance to a life of enslavement'.[55]

In addition, key questions remain as to how such creative practices and alternative imaginaries at the local level can be coordinated and extended to generate nationally transformative policies that benefit the many, and how transversal collaborations (and their associated imaginaries) can be deployed to aid this endeavour. Beyond particular initiatives, a key question remains how to use state institutional structures to reclaim, support and sustain more radical commons projects against threats of private appropriation.

More thought is also needed concerning how to transform the participatory character of many recent initiatives such as popular assemblies into more permanent organisational forms, as well as transforming the horizontal organisational logics and energy of spontaneity into institutional practices that are governed from below.[56] Debates about horizontality versus hierarchical organisational forms increasingly acknowledge that leaderlessness or pure horizontality is a myth and that some of the most powerful, successful movements in existence – such as the MST and the BKF – have quite traditional leadership structures. 'Leaderful' mobilisations have been suggested,[57] as exemplified by the Climate Games protests in Paris during the COP21 protests in 2015, and the global Break Free 2016 campaign that used direct action to attempt to shut down or disrupt major fossil-fuel sites in Germany, Spain, Turkey, Israel and Palestine, Indonesia, Philippines, Nigeria, South Africa, Canada, Brazil, Australia and the United States.[58]

Such governance from below will require a participatory politics that fosters relational, compositional and organisational power. Indeed, in their recent proclamations, the Zapatistas have called for combining critical thought with organised struggle from below, similar to Antonio Gramsci's philosophy of praxis, which calls for consciousness raising combined with political organisation.[59]

In addition, as I discussed in Chapter 6, there are compelling reasons at times for protestors to 'be as unknowable as the dark',[60] strategising in secure sequestered sites beyond surveillance technologies. In addition to the 'body politics' of non-violent direct action discussed in this book, activists might also consider the politics of disappearance or acting clandestinely to maintain independence given the dangers posed to protestors of increased police violence, CCTV surveillance, intelligence gathering and police infiltration of activist networks.[61] The targets of such action might still be physical, however. For example, the digital era of 'big data' requires telecommunications networks and computing infrastructures such as data centres and servers that are proliferating around the world, often using and transforming old industrial infrastructures such as factories.[62] These are important physical infrastructures for global capitalism. They comprise a potential tenth target of intervention – sites of information – that might be disrupted through virtual means, through computer hacking for example, or through more physical means such as occupations and blockades.

For example, the loosely associated international network of activists and hacktivists known as Anonymous have engaged in 'distributed denial of service' attacks against websites, which involve inundating computer servers with excessive quantities of data traffic so that they are unable to cope and shut down. A recent example of this occurred in 2010 when WikiLeaks began the release of 500,000 secret US diplomatic cables. The US Government coerced financial institutions including PayPal into cutting off service to WikiLeaks under threat of legal action. In response, Anonymous announced its support of WikiLeaks and waged cyber war against PayPal's transaction site, shutting it down.[63]

Finally, concerns over social reproduction – which underpin many of the sites of intervention that I have discussed in this book – must also be about the reproduction of our struggles.[64] Chantal Mouffe argues for the creation of a counter-hegemonic 'we' fashioned through chains of equivalence that embody a 'collective will across differences', waging a war of position against a multiplicity of 'nodal points of power that need to be targeted and transformed.[65] If key struggles of the twenty-first century are to establish what Mouffe calls a 'new' hegemony, then they will necessitate mobilising around the intersection of the key issues facing humanity (such as climate change, inequality and 'precarity') in order to fashion sustainable chains of equivalence. Fashioning innovative ways of global organising that are able to integrate localised frontline struggles with the modelling of alternative futures will also be required. The insurrectional must constantly strive to be intersectional.

The insurrectional will also require fashioning a 'we' that maintains a constant antagonism against capital and at times the state, and that disrupts the political order. New forms of struggle that combine confrontation, the creation of alternatives and new movement infrastructures and social relationships will need to evolve and adapt to ever-changing circumstances, generating a 'logic of overflow',[66] that is, spaces of mutation, experimentation and resistance. We might consider a reverse 'shock doctrine'[67] as a strategic intervention, so that, at times of political upheaval and economic disruption, we can organise collective attempts to create opportunities for movements and other forms of collective resistance, creativity and collaboration. We will need to target the sites of intervention discussed in this book, and no doubt new sites that will emerge as the current century unfolds.

Some of the most visible forms of intervention of the recent past have been the blockading of sites of circulation such as roads and ports and the occupation of city streets and squares, as seen, for example, in anti-roads protests and mobilisations by Occupy, Idle No More and Black Lives Matter. In part they are responding to economic changes over the past 30 years that have seen the increasing use of information technologies and the financialisation of the global economy through which capital accumulation has been increasingly focused on processes of circulation as well as production in order to generate surplus value. Global shipping, containerisation and the deregulation of transport create efficiencies more in circulation than in the organisation of production through distribution, inventory and exchange.[68]

As a result, protestors target infrastructures that enable circulation processes, each flow being an element in the overall reproduction of capitalism. For example, during 2016, mass mobilisations and assemblies of the Nuit Debout ('night on our feet') movement took place in town squares across France against economic inequality, unemployment and neoliberal labour reforms instituted by the French government. On 9

March 2016, some 500,000 took to the streets across France, actions accompanied by blockades of oil terminals by trade unions.[69]

We will all need to learn to become activists, acting and reflecting upon the world. This will require a range of practices. We will need to educate ourselves about everyday issues and conflicts and engage with social media to enable connections to be made between us and to facilitate knowledge-building. We will need to develop our relational power by finding like-minded people, and develop our organisational power by getting involved with local groups and campaigns. We will need to learn and develop new skills, drawing upon resources such as Beautiful Trouble and the Activists Handbook.[70] We will need to develop conscious activism, using our heads, hearts and hands, living the change we want to see in the world.

In fashioning radical futures a keen sense of geography will be critical. We will need to think spatially and strategically, combining all six spatial strategies outlined in Chapter 1 within and between struggles wherever possible. We must endeavour to not separate who we are from what we do and what we are becoming. The Zapatistas are a good example of this, having from the outset of their rebellion nurtured an autonomist indigenous sense of self, practice and potential. Like the popular assemblies of Occupy, we should not ask permission from the authorities before staging protests. We should just appear. It is also important that we do not just react to events. Rather, like the farmers of the BKF, we should be proactive, withdrawing our consent from the injustices of capitalism, and in so doing fashioning new ways of being and acting. Like the post-squares movements in Athens, and in climate justice struggles like the ZADs, we will need to intensify struggles for social reproduction and nurture varied arts for the self-defence of the commons. We will need to develop shared perceptions of unfolding situations and organise collectively across the planet. Deploying spatial strategies, we will need to develop our relational, compositional and organisational powers across multiple sites of intervention in order to refigure space. We will need to become space invaders.

Notes

CHAPTER 1

1. C. Tilly and S. Tarrow, *Contentious Politics* (Boulder, CO: Paradigm Publishers, 2007).
2. A. Melucci, *Nomads of the Present* (London: Radius, 1989); A. Melucci, *Challenging Codes: Collective Action in the Information Age* (Cambridge: Cambridge University Press, 1996).
3. D. Massey, *For Space* (London: Sage, 2005).
4. S.J. Smith, 'Society-Space', in P. Cloke, P. Crang and M. Goodwin (eds), *Introducing Human Geographies*, 2nd edn (London: Hodder Arnold, 2005), pp. 18–33.
5. Ibid.
6. E. McGurty, *Transforming Environmentalism: Warren County, PCBs, and the Origins of Environmental Justice* (New Brunswick, NJ: Rutgers University Press, 2007).
7. D. Della Porta and M. Diani (eds), *The Oxford Handbook of Social Movements* (Oxford: Oxford University Press, 2015).
8. For example, J. McCarthy and M. Zald, 'Resource Mobilization and Social Movements: A Partial Theory', *American Journal of Sociology* 82/6 (1977), 1212–1241.
9. A. Oberschall, *Social Conflict and Social Movements* (Englewood Cliffs, NJ: Prentice-Hall, 1973).
10. C. Tilly, *From Mobilisation to Revolution* (Reading, MA: Addison Wesley, 1978); Melucci, *Nomads of the Present*.
11. M. Castells, *The Power of Identity* (Oxford: Blackwell, 1997); M. Castells, *Networks of Outrage and Hope: Social Movements in the Internet Age* (Cambridge: Polity, 2012).
12. D.A. Snow, R.B. Rochford Jr., S.K. Worden and R.D. Benford, 'Frame Alignment Processes, Micromobilization, and Movement Participation', *American Sociological Review* 51 (1986), 464–481; R.D. Benford and D.A. Snow, 'Framing Processes and Social Movements: An Overview and Assessment', *Annual Review of Sociology* 26 (2000), 611–639.
13. J. Jasper, *The Art of Moral Protest* (Chicago: University of Chicago Press, 1997); J. Goodwin, J.A. Jasper and F. Polletta (eds), *Passionate Politics: Emotions and Social Movements* (Chicago: University of Chicago Press, 2001).
14. Tilly, *From Mobilisation to Revolution*; D. McAdam, *Political Process and the Development of Black Insurgency, 1930–1970* (Chicago: University of

Chicago Press, 1982); A. Touraine, *The Voice and the Eye: An Analysis of Social Movements* (Cambridge: Cambridge University Press, 1981).

15. P. Routledge, *Terrains of Resistance: Nonviolent Social Movements and the Contestation of Place in India* (Westport, CT: Praeger, 1993).

16. See W.J. Nicholls, 'Place, Relations, Networks: The Geographical Foundations of Social Movements', *Transactions of the Institute of British Geographers* 34/1 (2009), 78–93; W.J. Nicholls, B. Miller and J. Beaumont (eds), *Spaces of Contention: Spatialities and Social Movements* (Farnham: Ashgate Publishing, 2013).

17. J. Agnew, *Place and Politics: The Geographical Mediation of State and Society* (London: Allen & Unwin, 1987); D. Harvey, 'Militant Particularism and Global Ambition: The Conceptual Politics of Place, Space and Environment in the Work of Raymond Williams', *Social Text* 42 (1995), 69–98.

18. Routledge, *Terrains of Resistance*.

19. D. Martin, '"Place-Framing" as Place-Making: Constituting a Neighborhood for Organizing and Activism', *Annals of the Association of American Geographers* 93/3 (2003), 730–750.

20. P. Routledge, 'Going Globile: Spatiality, Embodiment and Media-tion in the Zapatista Insurgency', in S. Dalby and G. O'Tuathail (eds), *Rethinking Geopolitics* (London: Routledge, 1998), pp. 240–260.

21. B. Miller, *Geography and Social Movements: Comparing Antinuclear Activism in the Boston Area* (Minneapolis: University of Minnesota Press, 2000); S. Tarrow and D. McAdam, 'Scale Shift in Transnational Contention', in D. della Porta and S. Tarrow (eds), *Transnational Protest and Global Activism* (Boulder, CO: Rowman & Littlefield, 2005), pp. 121–150; W.J. Nicholls, 'The Geographies of Social Movements', *Geography Compass* 1/3 (2007), 607–622.

22. S. Tufts, 'Community Unionism in Canada and Labour's (Re)Organization of Space', *Antipode* 30/3 (1998), 227–250; J. Wills, 'Community Unionism and Trade Union Renewal in the UK: Moving Beyond the Fragments at Last?' *Transactions of the Institute of British Geographers* 26 (2001), 465–483; J. Wills, 'The Geography of Union Organising in Low-Paid Service Industries in the UK: Lessons from the T&G's Campaign to Unionise the Dorchester Hotel', *Antipode* 37 (2005), 139–159.

23. D. Massey, *Space Place and Gender* (Minneapolis: University of Minnesota Press, 1994), p. 156, original emphasis.

24. U. Oslender 'Fleshing Out the Geographies of Social Movements: Black Communities on the Colombian Pacific Coast and the Aquatic Space', *Political Geography* 23/8 (2004), 957–985; W. Wolford, 'This Land Is Ours Now: Spatial Imaginaries and the Struggle for Land in Brazil', *Annals of the Association of American Geographers*, 94/2 (2004), 409–424.

25. T. Cresswell, 'Place', in P. Cloke, P. Crang and M. Goodwin (eds), *Introducing Human Geographies*, 2nd edn (London: Hodder Arnold, 2005), pp. 485–494.

26. Miller, *Geography and Social Movements*.

27. McAdam, *Political Process*.

28. Nicholls, 'Geographies of Social Movements'.

29. A. Herod (ed.), *Organizing the Landscape: Geographical Perspectives of Labor Unionism* (Minneapolis: University of Minnesota Press, 1998); A. Herod, *Labor Geographies* (New York: Guildford Press, 2001).
30. For example, P. Waterman, *Globalization, Social Movements, and the New Internationalisms* (Washington: Mansell, 1998); J. Pickerill, *Cyberprotest: Environmental Activism On-Line* (Manchester: Manchester University Press, 2003).
31. H. Leitner, E. Sheppard and K.M. Sziarto, 'The Spatialities of Contentious Politics', *Transactions of the Institute of British Geographers* 33/2 (2008), 157–172.
32. A. Escobar, 'Culture Sits in Places: Reflections on Globalism and Subaltern Strategies of Localization', *Political Geography* 20/2 (2001), 139–174; F. Bosco, 'Place, Space, Networks, and the Sustainability of Collective Action: The *Madres de Plaza de Mayo*', *Global Networks* 1/4 (2001), 307–329.
33. N. Klein, *No Logo* (London: Flamingo, 2000).
34. N. Castree, 'Geographic Scale and Grassroots Internationalism: The Liverpool Dock Dispute 1995–1998', *Economic Geography* 76/3 (2000), 272–292.
35. A. Cumbers, P. Routledge and C. Nativel, 'The Entangled Geographies of Global Justice Networks', *Progress in Human Geography* 32/2 (2008), 179–197.
36. D. Featherstone, 'Spatialities of Transnational Resistance to Globalization: The Maps of Grievance of the Inter-Continental Caravan', *Transactions of the Institute of British Geographers* 28/4 (2003), 404–421.
37. J. Pickerill and P. Chatterton, ' Notes Towards Autonomous Geographies: Creation, Resistance and Self-Management as Survival Tactics', *Progress in Human Geography* 30 (2006), 730–746.
38. A. Feigenbaum, F. Frenzel and P. McCurdy, *Protest Camps* (London: Zed Books, 2013).
39. For example, H. Kurtz, 'Gender and Environmental Justice in Louisiana: Blurring the Boundaries of Public and Private Sphere', *Gender, Place and Culture* 14/4 (2007), 409–426; J. Hardy, W. Kozek and A. Stenning, 'In the Front Line: Women, Work and New Spaces of Labour Politics in Poland', *Gender, Place and Culture* 15/2 (2008), 99–116.
40. For example, F. Bosco, 'Emotions that Build Networks: Geographies of Human Rights Movements in Argentina and Beyond', *Tijdschrift voor Economische en Sociale Geografie* 98/5 (2007), 545–563; K. Askins, '"That's Just What I Do": Placing Emotion in Academic Activism', *Emotion, Space and Society* 2/1 (2009), 4–13; G. Brown and J. Pickerill, 'Space for Emotion in the Spaces of Activism', *Emotion, Space and Society* 2 (2009), 24–35; G. Pratt, 'Circulating Sadness: Witnessing Filipina Mothers' Stories of Family Separation', *Gender, Place and Culture* 16 (2009), 3–22; N. Clough, 'Emotion at the Center of Radical Politics: On the Affective Structures of Rebellion and Control', *Antipode* 44 (2012), 1667–1686.
41. P. Routledge, 'The Third Space as Critical Engagement' *Antipode* 28/4 (1996), 397–419.

42. See, for example, K. England, 'Getting Personal: Reflexivity, Positionality, and Feminist Research', *Professional Geographer* 46 (1994), 80–89; L.A. Staeheli and V.A. Lawson, 'Feminism, Praxis and Human Geography', *Geographical Analysis* 27 (1995), 321–338; G. Rose, 'Situating Knowledges: Positionality, Reflexivities and Other Tactics', *Progress in Human Geography* 21/3 (1997), 305–320; C. Katz, 'On the Grounds of Globalization: A Topography for Feminist Political Engagement', *Signs* 26/4 (2001), 1213–1229; L. Pulido, 'The Interior Life of Politics', *Ethics, Place and Environment* 6/1 (2003), 46–52; D. Massey, 'Geographies of Responsibility', *Geografiska Annaler: Series B, Human Geography* 86/1 (2004), 5–18.

43. P. Chatterton, D. Fuller and P. Routledge, 'Relating Action to Activism: Theoretical and Methodological Reflections', in S. Kindon, R. Pain and M. Kesby (eds), *Participatory Action Research Approaches and Methods: Connecting People, Participation and Place* (London: Routledge, 2008), pp. 216–222.

44. See www.peoplesgeography.org.

45. For example, D. Fuller and R. Kitchen (eds), *Critical Theory/Radical Praxis: Making a Difference Beyond the Academy?* (Victoria, BC: Praxis ePress, 2004); S. Koopman, 'Cutting through Topologies: Crossing Lines at the School of the Americas', *Antipode* 40/5 (2008), 825–847; P. Routledge, 'Acting in the Network: ANT and the Politics of Generating Associations', *Environment and Planning D: Society and Space* 26/2 (2008), 199–217.

46. For example, S. Kindon, R. Pain and M. Kesby (eds), *Participatory Action Research Approaches and Methods: Connecting People, Participation and Place* (London: Routledge, 2008); S. Wynne-Jones, P. North and P. Routledge, 'Practising Participatory Geographies: Potentials, Problems and Politics', *Area* 47/3 (2015), 218–221.

47. Pickerill and Chatterton, 'Notes Towards Autonomous Geographies', p. 730.

48. For example, Autonomous Geographies Collective, 'Beyond Scholar Activism: Making Strategic Interventions Inside and Outside the Neoliberal University' *ACME: An International E-Journal for Critical Geographies* 9/2 (2010), 245–275; S. Halvorsen, 'Militant Research against-and-beyond Itself: Critical Perspectives from the University and Occupy London', *Area* 47/4 (2015), 466–472; B. Russell, 'Beyond Activism/Academia: Militant Research and the Radical Climate and Climate Justice Movement(s)', *Area* 47/4, (2015), 222–229.

49. J. Martinez-Alier et al., 'Between Activism and Science: Grassroots Concepts for Sustainability Coined by Environmental Justice Organizations', *Journal of Political Ecology* 21 (2014), 19–60.

50. P. Routledge and K.D. Dreickson, 'Situated Solidarities and the Practice of Scholar-Activism', *Environment and Planning D: Society and Space* 33/3 (2015), 391–407; see also R. Nagar and S. Geiger, 'Reflexivity and Positionality in Feminist Fieldwork Revisited', in A. Tickell, E. Sheppard, J. Peck and T.J. Barnes (eds), *Politics and Practice in Economic Geography* (London: Sage, 2007), pp. 267–278.

51. K. Marx, *Capital*, Vol. 1 [1867] (New York: International Publishers, 1976).

52. D. Harvey, *The New Imperialism* (Oxford: Oxford University Press, 2003).
53. For indigenous people, these processes are entwined with ongoing forms of colonialism. See G.S. Coulthard, *Red Skin, White Masks: Rejecting the Colonial Politics of Recognition* (Minneapolis: University of Minnesota Press, 2014).
54. W. Brown, *Undoing the Demos: Neoliberalism's Stealth Revolution* (New York: Zone Books, 2015).
55. C. Mouffe, *Agonistics: Thinking the World Politically* (London: Verso Books, 2013); E. Swyngedouw, 'Apocalypse Forever? Post-Political Populism and the Spectre of Climate Change', *Theory, Culture and Society* 27/2-3 (2010), 213-232.
56. J. Rancière, *Dissensus: On Politics and Aesthetics* (London: Continuum, 2010), p. 36.
57. Ibid., p. 39.
58. E. Swyngedouw, 'Interrogating Post-Democratization: Reclaiming Egalitarian Political Spaces', *Political Geography* 30 (2011), 370-380.
59. Ibid.
60. G. Sharp, *The Politics of Nonviolent Action* (Boston: Porter Sargent, 1973); P. Ackerman, and C. Kruegler, *Strategic Nonviolent Conflict* (Westport, CT: Praeger, 1994).
61. J. Butler, *Notes on a Performative Theory of Assembly* (Cambridge, MA: Harvard University Press, 2015).
62. M. Zechner and B.R. Hansen, 'Building Power in a Crisis of Social Reproduction', *ROAR Magazine*, 2015 (available at: https://roarmagazine.org/magazine/building-power-crisis-social-reproduction/, accessed 4 January 2016).
63. M. Sitrin and D. Azzellini, *They Cannot Represent Us: Reinventing Democracy from Greece to Occupy* (London: Verso Books, 2014).
64. J. Sharp, P. Routledge, C. Philo and R. Paddison (eds), *Entanglements of Power: Geographies of Domination/Resistance* (London: Routledge, 2000).
65. For example, Leitner et al., 'Spatialities of Contentious Politics'; Nicholls, 'Place, Relations, Networks'; Nicholls et al., Spaces of Contention.
66. See J. Verson, 'Why We Need Cultural Activism' in Trapese Collective (eds), *Do It Yourself* (London: Pluto Press, 2007), pp. 171-186.
67. S.S. Karatasli, S. Kumral, B. Scully and S. Upadhyay, 'Class, Crisis and the 2011 Protest Wave: Cyclical and Secular Trends in Global Labor Unrest', in I. Wallestein, C. Chase-Dunn and C. Suter (eds), *Overcoming Global Inequalities* (London: Paradigm Publishers, 2015), pp. 184-200.
68. L. Laba, 'Solidarity (Poland)', in R.S. Powers, W.B. Vogele, C. Kruegler and R.M. McCarthy (eds), *Protest, Power, and Change* (New York: Garland Publishing, 1997), pp. 482-491.
69. T. Cresswell, *In Place/Out of Place* (Minneapolis: University of Minnesota Press, 1996).
70. G.T. Eskew, 'Civil Rights Movement (United States)', in R.S. Powers, W.B. Vogele, C. Kruegler and R.M. McCarthy (eds), *Protest, Power, and Change* (New York: Garland Publishing, 1997), pp. 86-89.

71. Feigenbaum et al., *Protest Camps*.
72. K. Mathiesen, K. 2015: 'Climate Change Activists Occupy Tate Modern's Turbine Hall', *Guardian*, 13 June 2015 (available at: https://www.theguardian.com/world/2015/jun/13/climate-change-activists-occupy-tate-moderns-turbine-hall, accessed 13 June 2015).
73. C. Mouffe, *Agonistics: Thinking the World Politically* (London: Verso, 2013).
74. A. Gramsci, *Selections from the Prison Notebooks* (New York: International Publishers, 1971), p. 243.
75. These are Venezuela, Mali, Senegal, Nepal, Ecuador, Bolivia, Nicaragua, Cuba, Dominican Republic, Peru, Argentina, Guatemala, Brazil, El Salvador and Indonesia. See C.M. Schiavoni, 'The Contested Terrain of Food Sovereignty Construction: Toward a Historical, Relational and Interactive Approach', *Journal of Peasant Studies* 44 (2017), 1–32.

CHAPTER 2

1. Material for this section is drawn in part from P. Routledge, 'Putting Politics in Its Place: Baliapal, India as a Terrain of Resistance', *Political Geography* 11/6 (1992), 588–611; and P. Routledge, *Terrains of Resistance: Nonviolent Social Movements and the Contestation of Place in India* (Westport, CT: Praeger, 1993).
2. S. Marglin and F. Marglin (eds), *Dominating Knowledge: Development, Culture and Resistance* (Oxford: Clarendon Press, 1990).
3. J.K. Russell, 'Blockade', in A. Boyd and D.O. Mitchell (eds), *Beautiful Trouble* (London: OR Books, 2012), pp. 14–17.
4. Material for this section is drawn in part from P. Routledge, 'Backstreets, Barricades, and Blackouts: Urban Terrains of Resistance in Nepal', *Society and Space* 12/5 (1994), 559–578; and P. Routledge, 'A Spatiality of Resistances: Theory and Practice in Nepal's Revolution of 1990', in S. Pile and M. Keith (eds), *Geographies of Resistance* (London: Routledge, 1997), pp. 68–86.
5. INSEC (Informal Sector Service Center), 'Nepal and its Electoral System' (Kathmandu: INSEC, 1991); INSEC, 'Identification of Bonded Labour in Nepal' (Kathmandu: INSEC, 1992).
6. INSEC, 'Nepal and Its Electoral System'; INSEC, 'Identification of Bonded Labour'; FOPHUR (Forum for Protection of Human Rights), 'Dawn of Democracy' (Kathmandu: FOPHUR, 1990).
7. R.I. Levy, *Mesocosm: Hinduism and the Organisation of a Traditional Newar City in Nepal* (Berkeley: University of California Press, 1990), pp. 188, 748n.
8. Ibid., p. 188.
9. Ibid., pp. 182–186.
10. Ibid., pp. 182–186.
11. FOPHUR, 'Dawn of Democracy'.
12. P. Routledge, 'Nineteen Days in April: Urban Protest and Democracy in Nepal', *Urban Studies* 47/6 (2010), 1279–1299.

13. D. Gregory, *The Colonial Present* (Oxford: Wiley-Blackwell, 2004); M. Lecoquirre, 'Holding On to Place: Spatialities of Resistance in Israel and Palestine, the Cases of Hebron, Silwan and al-Araqib', PhD thesis (Florence: Department of Political Science, European University Institute, 2016).

14. A. Bayat, 'The Art of Presence', *ISIM Newsletter*, 14 June 2004 (available at: https://openaccess.leidenuniv.nl/bitstream/handle/1887/16951/ISIM_14_The_Art_of_Presence.pdf?sequence=1. pp. 5, accessed 24 July 2016).

15. The Invisible Committee, *For Our Friends* (South Pasadena, CA: Semiotext(e), 2015).

16. Lecoquirre, 'Holding On to Place'.

17. Ibid.

18. Ibid.

19. Ibid.

20. Ibid.

CHAPTER 3

1. S. Halvorsen, 'Taking Space: Moments of Rupture and Everyday Life in Occupy London', *Antipode* 47/2 (2015), 401–417.

2. R. Zibechi, *Dispersing Power: Social Movements as Anti-State Forces* (Edinburgh: AK Press, 2010); R. Zibechi, *Territories in Resistance: A Cartography of Latin American Social Movements* (Edinburgh: AK Press, 2012).

3. Zibechi, *Territories in Resistance*, p. 19.

4. See also A. Escobar, *Territories of Difference: Place, Movements, Life* (Durham, NC: Duke University Press, 2008).

5. G.S. Coulthard, *Red Skin, White Masks: Rejecting the Colonial Politics of Recognition* (Minneapolis: University of Minnesota Press, 2014).

6. Zibechi, *Territories in Resistance*, pp. 7–8.

7. G. Caffentzis, 'In the Desert of Cities: Notes on the Occupy Movement in the USA', paper presented at 'The Tragedy of the Market: From Crisis to Commons', Vancouver, BC, 8 January 2012 (available at: http://www.reclamationsjournal.org/blog/?p=505, accessed 26 June 2012); P. Chatterton, D. Featherstone and P. Routledge, 'Articulating Climate Justice in Copenhagen: Antagonism, the Commons and Solidarity', *Antipode* 45/3 (2013), 602–620.

8. Material for this section is drawn in part from P. Routledge, 'The Imagineering of Resistance: Pollok Free State and the Practice of Postmodern Politics', *Transactions of the Institute of British Geographers* 22 (1997), 359–376.

9. S. Halvorsen, 'Beyond the Network? Occupy London and the Global Movement', *Social Movement Studies* 11/3–4 (2012), 427–433.

10. A. Feigenbaum, F. Frenzel and P. McCurdy, *Protest Camps* (London: Zed Books, 2013).

11. Material for this section is drawn in part from P. Routledge, 'Acting in the Network: ANT and the Politics of Generating Associations', *Environment and Planning D: Society and Space* 26/2 (2008), 199–217; and P. Routledge,

'Territorializing Movement: The Politics of Land Occupation in Bangladesh' *Transactions of the Institute of British Geographers* 40/4 (2015), 445–463.

12. C. Parenti, *Tropic of Chaos: Climate Change and the New Geography of Violence* (New York: Nation Books, 2011).

13. IPCC (Intergovernmental Panel on Climate Change), 'Climate Change 2007: Synthesis Report' (available at: https://www.ipcc.ch/ pdf/assessment-report/ar4/syr/ar4_syr.pdf, accessed 24 January 2017).

14. M.F. Karim and N. Mimura, 'Impacts of Climate Change and Sea Level Rise on Cyclonic Storm Surge Floods in Bangladesh', *Global Environmental Change* 18 (2008), 490–500.

15. B.K. Paul and S. Dutt, 'Hazard Warnings and Responses to Evacuation Orders: The Case of Bangladesh's Cyclone Sidr', *Geographical Review* 100/3 (2010), 336–355.

16. R. Brouwer, S. Akter, L. Brander and E. Haque, 'Socioeconomic Vulnerability and Adaptation to Environmental Risk: A Case Study of Climate Change and Flooding in Bangladesh', *Risk Analysis* 27/2 (2007), 313–326.

17. S. Dasgupta, M. Huq, Z.H. Khan, M.S. Masud, M. Ahmed, N. Mukherjee and K. Pandey, 'Climate Proofing Infrastructure in Bangladesh: The Incremental Cost of Limiting Future Flood Damage', *Journal of Environment and Development* 20/2 (2011), 167–190.

18. J. Devine, 'Ethnography of a Policy Process: A Case Study of Land Distribution in Bangladesh', *Public Administration and Development* 22/5 (2002), 403–422; J. Devine, 'NGOs, Politics and Grassroots Mobilisation: Evidence from Bangladesh', *Journal of South Asian Development* 1/1 (2006), 77–99.

19. K. Murshid, 'Implications of Agricultural Policy Reforms on Rural Food Security and Poverty', undated (available at: http://www.saprin.org/ bangladesh/research/ban_agri_policy.pdf, accessed 24 January 2017).

20. See also D. Nally, 'The Biopolitics of Food Production', *Transactions of the Institute of British Geographers* 36 (2011), 37–53.

21. M. Hossain, 'Dynamics of Poverty in Rural Bangladesh 1988–2007: An Analysis of Household Level Panel Data', unpublished paper presented at the conference 'Employment Growth and Poverty Reduction in Developing Countries', Political Economy Research Institute, University of Massachusetts, Amherst, 27–28 March 2009 (available at: http://wwwperiumassedu/ fileadmin/pdf//Hossain_Bangladeshdoc, accessed 13 August 2015).

22. W. Wolford, 'This Land Is Ours Now: Spatial Imaginaries and the Struggle for Land in Brazil', *Annals of the Association of American Geographers,* 94/2 (2004), 409–424.

23. B. Baletti, T.M. Johnson and W. Wolford, '"Late Mobilization": Transnational Peasant Networks and Grassroots Organizing in Brazil and South Africa', *Journal of Agrarian Change* 8/2–3 (2008), 290–314.

24. Quoted in Routledge, 'Territorializing Movement', p. 451.

25. B. Alam, *The Struggle* (Calcutta), October 2008, p. 4.

26. H. Wittman, 'Reworking the Metabolic Rift: La Via Campesina, Agrarian Citizenship and Food Sovereignty', *Journal of Peasant Studies* 36/4 (2009), 805–826.

27. R. Patel, 'What Does Food Sovereignty Look Like?', *Journal of Peasant Studies* 36/3 (2009), 663–673; P.M. Rosset, B.M. Sosa, A.M.R. Jaime and D.R.A. Lozano, 'The Campesino-to-Campesino Agroecology Movement of ANAP in Cuba: Social Process Methodology in the Construction of Sustainable Peasant Agriculture and Food Sovereignty', *Journal of Peasant Studies* 38/1 (2011), 161–191.

28. Quoted in Routledge, 'Acting in the Network', p. 208.

29. S.S. Karatasli, S. Kumral, B. Scully and S. Upadhyay, 'Class, Crisis and the 2011 Protest Wave: Cyclical and Secular Trends in Global Labor Unrest', in I. Wallestein, C. Chase-Dunn and C. Suter (eds), *Overcoming Global Inequalities* (London: Paradigm Publishers, 2015), pp. 184–200; J. Clover, *Riot. Strike. Riot: The New Era of Uprisings* (London: Verso, 2016).

30. J. Shenker, *The Egyptians: A Radical Story* (London: Allen Lane, 2016).

31. Source: https://en.wikipedia.org/wiki/Tahrir_Square (accessed 27 September 2016).

32. Feigenbaum at al., *Protest Camps*.

33. Shenker, *The Egyptians*.

34. J. Juris, 'Reflections on #Occupy Everywhere: Social Media, Public Space, and Emerging Logics of Aggregation', *American Ethnologist* 39/2 (2012), 259–279.

35. P. Marcuse, 'Keeping Space in Its Place in the Occupy Movements', *Progressive Planning* 191 (Spring 2012), 15–16.

36. Juris, 'Reflections on #Occupy Everywhere'.

37. J. Pickerill and J. Krinsky, 'Why Does Occupy Matter?' *Social Movement Studies* 11/3–4 (2012), 279–287; M. Kaika and L. Karaliotias, 'The Spatialization of Democratic Politics: Insights from Indignant Squares', *European Urban and Regional Studies* 23/4 (2014), 1–54.

38. Clover, *Riot*.

39. Pickerill and Krinsky, 'Why Does Occupy Matter?'; J.L. Hammond, 'The Significance of Space in Occupy Wall Street', *Interface: A Journal for and about Social Movements* 5/2 (2013), 499–524.

40. Pickerill and Krinsky, 'Why Does Occupy Matter?'

41. S. Federici, *Revolution at Point Zero: Housework, Reproduction and Feminist Struggle* (Oakland, CA: PM Press, 2012), p. 145.

42. Caffentzis, 'In the Desert of Cities'.

43. J. Butler and A. Athanasiou, *Dispossession: The Performative in the Political* (Cambridge: Polity Press, 2013).

44. Pickerill and Krinsky, 'Why Does Occupy Matter?'; Hammond, 'The Significance of Space'.

45. M. Sitrin and D. Azzellini, *They Cannot Represent Us: Reinventing Democracy from Greece to Occupy* (London: Verso Books, 2014).

46. Pickerill and Krinsky, 'Why Does Occupy Matter?'

47. G. Jobin-Leeds and AgitArte, *When We Fight We Win: Twenty-First-Century Social Movements and the Activists that are Transforming our World* (New York: New Press, 2016).

48. Caffentzis, 'In the Desert of Cities'; Pickerill and Krinsky, 'Why Does Occupy Matter?'

49. K. Aronoff, 'Occupy Had Its Shortcomings, but It Was Hardly Forgettable', *Waging Nonviolence*, 10 June 2015 (available at: http://wagingnonviolence.org/2015/06/occupy-shortcomings-hardly-forgetable/, accessed 24 January 2017).

50. Juris, 'Reflections on #Occupy Everywhere'.

51. Marcuse, 'Keeping Space in Its Place'.

52. J.M. Smucker, 'Can Prefigurative Politics Replace Political Strategy?' *Berkeley Journal of Sociology* 58 (2014) (available at: http://berkeleyjournal.org/2014/10/can-prefigurative-politics-replace-political-strategy/, accessed 24 January 2017).

53. Sitrin and Azzellini, *They Cannot Represent Us*.

54. Ibid.

55. The Invisible Committee, *For Our Friends* (South Pasadena, CA: Semiotext(e), 2015).

56. A. Arampatzi, 'The Spatiality of Counter-Austerity Politics in Athens, Greece: Emergent "Urban Solidarity Spaces"' *Urban Studies* (in press).

57. Sitrin and Azzellini, *They Cannot Represent Us*.

CHAPTER 4

1. On the pack, see E. Canetti, *Crowds and Power* (Harmondsworth: Penguin, 1962); on the swarm, see K. Ross, *The Emergence of Social Space: Rimbaud and the Paris Commune* (Basingstoke: Macmillan, 1988).

2. Canetti, *Crowds and Power* p. 85.

3. G. Deleuze and F. Guattari, *A Thousand Plateaus* (Minneapolis: University of Minnesota Press, 1987).

4. R. Williams, *Marxism and Literature* (Oxford: Oxford University Press, 1977), pp. 128–135.

5. Material for this section is drawn in part from P. Routledge, 'Space, Mobility, and Collective Action: India's Naxalite Movement', *Environment and Planning A* 29 (1997), 2165–2189.

6. From 'Peasant's Song' by Nitya Sen, in S. Banerjee, *Thema Book of Naxalite Poetry* (Calcutta: Thema, 1987), p. 38.

7. S. Banerjee, *India's Simmering Revolution* (London: Zed Press, 1984).

8. Ibid., pp. 4–5.

9. P. Bardhan, *The Political Economy of Development in India* (Delhi: Oxford University Press, 1985).

10. In Maoist communist ideology, peasants are considered a revolutionary force, and Maoists have initiated rural-based mass movements to fashion bases of liberated space in the countryside. Over time such bases are

intended to unite and surround the cities, overthrowing the ruling regime and seizing political power.

11. Banerjee, *India's Simmering Revolution*; B. Dasgupta, *The Naxalite Movement* (Bombay: Allied, 1974); E. Duyker, *Tribal Guerrillas* (Delhi: Oxford University Press, 1987).

12. Dasgupta, *Naxalite Movement*.

13. C. Mazumdar, 'A Few Words about Guerrilla Actions', 1970 (available at https://cpiml.org/charu-mazumdar-collected-writings/formation-of-communist-party-of-india-marxist-leninist-22-april-1969/a-few-words-about-guerrilla-actions, accessed 22 July 2016).

14. S. Ghosh, *The Naxalite Movement: A Maoist Experiement* (Calcutta: Firma, 1974).

15. Duyker, *Tribal Guerrillas*.

16. Ibid.

17. B. Joshi, 'India and the Backdoor Emergency', *South Asia Bulletin* 5/2 (1985), 14–24.

18. Banerjee, *India's Simmering Revolution*, pp. 186–189.

19. C. Parenti, *Tropic of Chaos: Climate Change and the New Geography of Violence* (New York: Nation Books, 2011).

20. A. Roy, 'Walking with the Comrades', *Outlook*, 29 March 2010 (available at: http://www.outlookindia.com/magazine/story/walking-with-the-comrades/264738, accessed 18 May 2016).

21. K.D. Derickson, 'The Racial State and Resistance in Ferguson and Beyond', *Urban Studies* 53/11 (2016), 2223–2237.

22. D. Cowen and N. Lewis, 'Anti-Blackness and Urban Geopolitical Economy: Reflections on Ferguson and the Suburbanization of the "Internal Colony"', *Environment and Planning D: Society and Space* 24/2 (2016), 159–163.

23. Ibid.

24. J. Clover, *Riot. Strike. Riot: The New Era of Uprisings* (London: Verso, 2016); R.W. Gilmore, *Golden Gulag: Prisons, Surplus, Crisis and Opposition in Globalizing California* (Berkeley: University of California Press, 2007).

25. K. Davidson, 'These 7 Household Names Make a Killing Off of the Prison Industrial Complex', 2015 (available at: http://usuncut.com/class-war/these-7-household-names-make-a-killing-off-of-the-prison-industrial-complex/, accessed 3 March 2016).

26. G. Jobin-Leeds and AgitArte, *When We Fight We Win: Twenty-First-Century Social Movements and the Activists that are Transforming our World* (New York: New Press, 2016).

27. Quoted from: http://blacklivesmatter.com/about/ (accessed 12 August 2016).

28. Jobin-Leeds and AgitArte, *When We Fight We Win*.

29. Clover, *Riot*, p. 182.

30. J. Swaine, 'Ferguson Protests: State of Emergency Declared after Violent Night', *Guardian*, 11 August 2015 (available at: https://www.theguardian.com/us-news/2015/aug/10/ferguson-protests-st-louis-state-of-emergency, accessed 12 June 2016).

31. Derickson, 'The Racial State'.
32. M. McIntee, 'Mall of America Protest a 'Decoy' Says Black Lives Matter', *The Uptake*, 23 December 2015 (available at: http://theuptake.org/2015/12/23/mall-of-america-protest-a-decoy-says-black-lives-matter/, accessed 12 June 2016).
33. Cowen and Lewis, 'Anti-Blackness and Urban Geopolitical Economy'.
34. H. Siddique, 'Anti-Racism Protests Block Three Cities', *Guardian*, 6 June 2016, p. 5.
35. J. Bruinsma, 'Cop Watch: Protecting Neighbourhoods from the Police', *ROAR Magazine*, 29 August 2015 (available at: https://roarmag.org/essays/cop-watch-police-us-murder-blacks-racism/, accessed 14 June 2016).
36. R. Kaulingfreks and S. Warren, 'SWARM: Flash Mobs, Mobile Clubbing and the City', *Culture and Organization* 16/ 3 (2010), 211–227.
37. D. Mitchell and A. Boyd, 'Flash Mobs', in A. Boyd and D. Mitchell (eds), *Beautiful Trouble: A Toolbox for Revolution* (London: OR Books, 2012), pp. 46–47.
38. See: www.ukuncut.org.uk/about/ (accessed 7 July 2016).
39. G.S. Coulthard, *Red Skin, White Masks: Rejecting the Colonial Politics of Recognition* (Minneapolis: University of Minnesota Press, 2014).
40. Jobin-Leeds and AgitArte, *When We Fight We Win*.
41. See: http://beautifultrouble.org/case/idle-dance-flash-mob/ (accessed 10 July 2016).
42. See: http://www.idlenomore.ca (accessed 12 August 2016).
43. L.J. Wood, 'Idle No More: Facebook and Diffusion', *Social Movement Studies* 14/5 (2015), 615–621.
44. Coulthard, *Red Skin, White Masks*..
45. See: http://beautifultrouble.org/case/idle-dance-flash-mob/ (accessed 6 May 2016).
46. S. Hind, 'Maps, Kettles and Inflatable Cobblestones: The Art of Playful Disruption in the City', *Media Fields Journal*, 21 August 2015 (available at: http://www.mediafieldsjournal.org/tactical-frivolity-and-disobed/, accessed 2 February 2016).
47. Ibid.
48. See: www.labofii.net (accessed 12 January 2016).
49. See: www.climategames.net (accessed 12 January 2016).
50. The Laboratory of Insurrectionary Imagination, 'Hacking the COP: The Climate Games in Paris 2015', *EJOLT*, 2015 (available at: http://www.ejolt.org/2015/09/refocusing-resistance-climate-justice-coping-coping-beyond-paris/, accessed 28 January 2017).
51. N. Lampert, 'D12 Paris Climate Demos Recap: Art and Activism', *Justseeds*, 29 December 2015 (available at: http://justseeds.org/d12-paris-climate-demos-recap-art-and-activism/, accessed 22 January 2016); K. Buckland, 'Cultural Shifts in the Climate Justice Movement', *Resilience*, 29 March 2016 (available at: http://www.resilience.org/stories/2016-03-29/cultural-shifts-in-the-climate-justice-movement, accessed 22 January 2016).

CHAPTER 5

1. J. Butler, *Notes Toward a Performative Theory of Assembly* (Cambridge, MA: Harvard University Press, 2015).
2. A. Melucci, *Nomads of the Present* (London: Radius, 1989).
3. D. Canning and P. Reinsborough, 'Think Narratively', in A. Boyd and D. Mitchell (eds), *Beautiful Trouble: A Toolbox for Revolution* (London: OR Books, 2012), pp. 186–187.
4. A. Feigenbaum, F. Frenzel and P. McCurdy, *Protest Camps* (London: Zed Books, 2013), p. 112.
5. Ibid., p. 112.
6. See: www.indymedia.org.
7. For example, see T. Olesen, *International Zapatismo* (London: Zed Books, 2005); J. Juris, 'Reflections on #Occupy Everywhere: Social Media, Public Space, and Emerging Logics of Aggregation', *American Ethnologist* 39/2 (2012), 259–279.
8. Material for this section is drawn in part from P. Routledge, 'Voices of the Dammed: Discursive Resistance amidst Erasure in the Narmada Valley, India', *Political Geography* 22/3 (2003), 243–270.
9. See A. Roy, *The Greater Common Good* (Bombay: India Book Distributors, 1999); S. Sangvai, *The River and Life* (Mumbai: Earthcare Books, 2000).
10. R. Dwivedi, 'Displacement, Risks and Resistance: Local Perceptions and Actions in the Sardar Sarovar', *Development and Change* 30 (1999), 43–75.
11. For example, International Rivers Network (IRN), 'Sardar Sarovar Project: An Overview' (Berkeley, CA: IRN, 1994); P. McCully, *Silenced Rivers: The Ecology and Politics of Large Dams* (London: Zed Books, 1996).
12. A. Baviskar, 'Written on the Body, Written on the Land: Violence and Environmental Struggles in Central India in 2001', in N.L. Peluso and M. Watts (eds) *Violent Environments* (Ithaca, NY: Cornell University Press, 2001), pp. 354–379.
13. K. Warren, 'Narrating Cultural Resurgence: Genre and Self-Representation for Pan-Mayan Writers', in D. Reed-Danahay (ed.), *Auto/Ethnography: Rewriting the Self and the Social* (Oxford: Berg, 1997), pp. 21–45, quote from p. 22.
14. Quoted in Routledge, 'Voices of the Dammed', pp. 261–262.
15. A. Gandhi, 'State (Under)Development, Transnational Activism and Tribal Resistance in India's Narmada Valley', MA thesis (Montreal: Department of Anthropology, McGill University, 2001).
16. Quoted in Routledge, 'Voices of the Dammed', p. 262.
17. Material for this section is drawn in part from P. Routledge, 'Going Globile: Spatiality, Embodiment and Media-tion in the Zapatista Insurgency', in S. Dalby and G. O'Tuathail (eds), *Rethinking Geopolitics* (London: Routledge, 1998), pp. 240–260.
18. EZLN (Ejército Zapatista de Liberación Nacional), 'Chiapas: The Southeast in Two Winds, a Storm and a Prophecy', *Anderson Valley Advertiser*, 42/31 (1994), pp.1–5, quote from p. 5.

19. J. Ross, *Rebellion from the Roots* (Monroe, ME: Common Courage Press, 1995).

20. EZLN, 'Declaration of the Lacandon Jungle: Today We Say "Enough"', *Anderson Valley Advertiser* 42/31 (1994), pp. 5–6.

21. Ross, *Rebellion from the Roots*.

22. EZLN communiqué, quoted in G.A. Collier and E.L. Quaratiello, *Basta! Land and the Zapatista Rebellion in Chiapas* (Oakland, CA: Institute for Food and Development Policy, 1994), p. 64.

23. EZLN, 'Zapata Will Not Die by Arrogant Decree', *Anderson Valley Advertiser* 42/31 (1994), p. 16.

24. Quoted in T. Golden, 'Mexican Rebel Leader Sees No Quick Settlement', *New York Times*, 20 February 1994 (available at: http://www.nytimes.com/1994/02/20/world/mexican-rebel-leader-sees-no-quick-settlement.html?pagewanted=all, accessed 1 April 2016).

25. See G. O'Tuathail, 'Emerging Markets and Other Simulations: Mexico, the Chiapas Revolt and the Geo-Financial Panopticon', *Ecumene* 4/3 (1997), 300–317.

26. S. Marcos, 'The Zapatista Women's Revolutionary Law as It Is Lived Today', *Open Democracy*, 22 July 2014 (available at: https://www.opendemocracy.net/sylvia-marcos/zapatista-women's-revolutionary-law-as-it-is-lived-today, accessed 3 March 2016).

27. R. Zibechi, *Territories in Resistance: A Cartography of Latin American Social Movements* (Edinburgh: AK Press, 2012).

28. A. Ewen, 'Mexico: The Crisis of Identity', *Akwe:kon* 11/2 (1994), 28–40.

29. Marcos, 'Zapatista Women's Revolutionary Law'.

30. EZLN, *Critical Thought in the Face of the Capitalist Hydra, 1: Contributions of the Sixth Commission of the EZLN* (Durham, NC: Paper Boat Press, 2016).

31. Quoted from: http://enlacezapatista.ezln.org.mx (accessed 6 April 2016).

32. See: www.adbusters.org.

33. M. Dery, 'Culture Jamming: Hacking, Slashing and Sniping in the Empire of Signs', Open Magazine Pamphlet Series (Vancouver, BC: Open Media, 1993) (available at: http://project.cyberpunk.ru/idb/culture, accessed 18 May 2016); N. Klein, *No Logo* (London: Flamingo, 2000).

34. See: www.brandalism.org.uk/gallery (accessed 22 January 2016).

35. T. Milstein, and A. Pulos, 'Culture Jam Pedagogy and Practice: Relocating Culture by Staying on One's Toes', *Communication, Culture and Critique* 8/3 (2015), 395–413.

36. J. Perini, 'Art as Intervention: A Guide to Today's Radical Art Practices', in Team Colors Collective, *Uses of a Whirlwind: Movement, Movements, and Contemporary Radical Currents in the United States* (New York: AK Press, 2010).

37. S. Hind, 'Maps, Kettles and Inflatable Cobblestones: The Art of Playful Disruption in the City', *Media Fields Journal*, 21 August 2015 (available at: http://www.mediafieldsjournal.org/tactical-frivolity-and-disobed/, accessed 20 June 2016).

38. Quoted in A. Escobar, 'Culture, Economics, and Politics in Latin American Social Movements Theory and Research', in A. Escobar and S.E. Alvarez (eds), *The Making of Social Movements in Latin America* (Boulder, CO: Westview Press, 1992), pp. 62–85, quote from p. 75.

39. W.L. Bennett and A. Segerberg, 'Communication in Movements', in D. Della Porta and M. Diani (eds), *The Oxford Handbook of Social Movements* (Oxford: Oxford University Press, 2015), p. 275.

CHAPTER 6

1. S. Marcos, 'Tomorrow Begins Today', in Notes from Nowhere (eds), *We are Everywhere: The Irresistible Rise of Global Anti-Capitalism* (London: Verso, 2003), pp. 8–13.

2. G.L. Ribeiro, 'Cybercultural Politics: Political Activism at a Distance in a Transnational World', in S.E. Alvarez, E. Dagni and A. Escobar (eds), *Cultures of Politics, Politics of Cultures* (Boulder, CO: Westview Press, 1998), pp. 325–352.

3. J. Butler, *Notes Toward a Performative Theory of Assembly* (Cambridge, MA: Harvard University Press, 2015).

4. Ibid., 92.

5. For example, see P. Routledge, *Terrains of Resistance: Nonviolent Social Movements and the Contestation of Place in India* (Westport, CT: Praeger, 1993).

6. P. Routledge, 'Convergence Space: Process Geographies of Grassroots Globalisation Networks', *Transactions of the Institute of British Geographers* 28/3 (2003), 333–349; P. Routledge and A. Cumbers, *Global Justice Networks: Geographies of Transnational Solidarity* (Manchester: Manchester University Press, 2009).

7. A. Escobar, 'Culture Sits in Places: Reflections on Globalism and Subaltern Strategies of Localization', *Political Geography* 20/2 (2001), 139–174.

8. See J. Sharp, P. Routledge, C. Philo and R. Paddison (eds), *Entanglements of Power Geographies of Domination/Resistance* (London: Routledge, 2000).

9. Material for this section is drawn in part from Routledge and Cumbers, *Global Justice Networks*, pp. 103–138.

10. T. Wallgren, 'Political Semantics of "Globalization": A Brief Note', *Development* 41/2 (1998), 30–32.

11. Solidarities across borders are not a new phenomena. During the nineteenth century, international alliances were established in the anti-slavery movement and the communists and anarchist internationals. In the twentieth century, internationalism has been present in the campaign for women's suffrage, the International Brigades during the Spanish Civil War in the 1930s, Cuba since the 1960s and Nicaragua in the 1980s, trade union activism and in the anti-nuclear movement. However, contemporary international alliances are characterised by the means, speed and intensity of communication between the various groups involved.

12. See: www.agp.org

13. J. Juris, *Networking Futures: The Movements against Corporate Globalization* (Durham, NC: Duke University Press, 2008).
14. See: https://www.nadir.org/nadir/initiativ/agp/ (accessed 12 October 2015).
15. F. Bosco, 'Place, Space, Networks, and the Sustainability of Collective Action: The *Madres de Plaza de Mayo*', *Global Networks* 1/4 (2001), 307–329.
16. N. Klein, *Fences and Windows* (London: Flamingo, 2002).
17. The G8 is an inter-governmental forum for the world's most developed economies, and currently comprises the United States, Canada, the UK, France, Germany, Italy, Japan and Russia.
18. Quoted in Routledge and Cumbers, *Global Justice Networks*, p. 113.
19. D.J. Featherstone, *Solidarity: Hidden Histories and Geographies of Internationalism* (London: Zed Books, 2012).
20. Quoted in Routledge and Cumbers, *Global Justice Networks*, p. 114.
21. These caravans have a certain historical precedent in the solidarity convoys that took North American activists to Nicaragua, El Salvador and Guatemala during the 1980s. These convoys brought humanitarian aid to communities and articulated opposition to US government policies in the region – particularly US support for the military juntas in El Salvador and Guatemala, and for the Contra war attempting to destabilise the Sandinista revolution in Nicaragua.
22. B.D. Missingham, *The Assembly of the Poor in Thailand* (Chiang Mai: Silkworm Books, 2003).
23. J. King, 'The Packet Gang', *Mute* 1/27 (2004) (available at: http://www.metamute.org/editorial/articles/packet-gang, accessed 30 January 2017).
24. Quoted in Routledge and Cumbers, *Global Justice Networks*, p. 117.
25. Quoted in ibid., p. 129.
26. Material for this section is drawn in part from P. Routledge, 'Engendering Gramsci: Gender, the Philosophy of Praxis and Spaces of Encounter in the Climate Caravan, Bangladesh', *Antipode* 47/5 (2015), 1321–1345.
27. Quoted in Routledge, 'Engendering Gramsci', p. 1321.
28. Participants came from India (Andhra Pradesh Vyavasaya Vruthidarula Union; Karnataka State Farmer's Association; Institute for Motivating Self Employment), Nepal (All Nepal Peasants' Federation; All Nepal Peasants' Federation (Revolutionary); All Nepal Women's Association; General Federation of Nepalese Trade Unions; Jagaran Nepal), Pakistan (Anjuman Muzareen Punjab/Punjab Tenants Association), Sri Lanka (Movement for National Land and Agricultural Reform; National Socialist Party) and the Philippines (Kilusang Magbubukid ng Pilipinas/Peasant Movement of the Philippines), as well as activists from La Via Campesina (South Asia), the UK, Germany and Australia
29. E. Barvosa-Carter, 'Multiple Identity and Coalition-Building: How Identity Differences within Us Enable Radical Alliances among Us', in J. Bystydzienski and S. Schacht (eds), *Forging Radical Alliances across Difference: Coalition Politics for the New Millennium* (Lanham, MD: Rowman & Littlefield, 2001), pp. 21–34.

30. D. Snowand R. Benford, 'Alternative Types of Cross-National Diffusion in the Social Movement Arena', in D. della Porta and H. Kriesi (eds), *Social Movements in a Globalising World* (New York: St Martin's Press, 1999), pp. 23–39.

31. C. Katz, 'On the Grounds of Globalization: A Topography for Feminist Political Engagement', *Signs* 26/4 (2001), 1213–1229.

32. P. Routledge, 'Acting in the Network: ANT and the Politics of Generating Associations', *Environment and Planning D: Society and Space* 26/2 (2008), 199–217.

33. Quoted in Routledge, 'Engendering Gramsci', p. 1331.

34. J. Bandy and J. Smith (eds), *Coalitions Across Borders* (Lanham, MD: Rowman & Littlefield, 2005).

35. Quoted in P. Routledge and K. Derickson, 'Situated Solidarities and the Practice of Scholar-Activism', *Environment and Planning D: Society and Space* 33/3 (2015), 391–407, quote from p. 401.

36. D.J. Featherstone, 'Spatialities of Transnational Resistance to Globalization: The Maps of Grievance of the Inter-Continental Caravan', *Transaction of the Institute of British Geographers* 28/4 (2003), 404–421.

37. Quoted in Routledge, 'Engendering Gramsci', p. 1335.

38. Routledge, 'Acting in the Network'.

39. D. Massey, *Space Place and Gender* (Minneapolis: University of Minnesota Press, 1994).

40. T. Mertes, 'Grass-Roots Globalism', *New Left Review* 17 (2002), 101–110.

CHAPTER 7

1. A. Gramsci, *Selections from the Prison Notebooks* (New York: International Publishers, 1971).

2. P. Chatterton, D. Fuller and P. Routledge, 'Relating Action to Activism: Theoretical and Methodological Reflections', in S. Kindon, R. Pain and M. Kesby (eds), *Participatory Action Research Approaches and Methods: Connecting People, Participation and Place* (London: Routledge, 2008), pp. 216–222.

3. S. Ahmed, *The Cultural Politics of Emotions* (New York: Routledge, 2004).

4. For example, see J. Goodwin, J.A. Jasper and F. Polletta (eds), *Passionate Politics: Emotions and Social Movements* (Chicago: University of Chicago Press, 2001); J. Jasper, 'The Emotions of Protest: Affective and Reactive Emotions in and around Social Movements', *Sociological Forum* 13 (1998), 397–424; J. Juris, 'Performing Politics: Image, Embodiment, and Emotive Solidarity during Anti-Corporate Globalization Protests', *Ethnography* 9/1 (2008), 61–97.

5. S. Wulff, M. Bernsteinand V. Taylor, 'New Theoretical Directions from the Study of Gender and Sexuality Movements', in D. della Porta and M. Diani (eds), *The Oxford Handbook of Social Movements* (Oxford: Oxford University Press, 2015), pp. 108–130.

6. J. Reger, 'Organizational "Emotion Work" through Consciousness-Raising: An Analysis of a Feminist Organization', *Qualitative Sociology* 27/2 (2004), 205–222.

7. See G. O'Tuathail, '"Just Out Looking For a Fight": American Affect and the Invasion of Iraq', *Antipode* 35 (2003), 856–870; U. Oslender, 'Spaces of Terror and Fear on Colombia's Pacific Coast', in D. Gregory and A. Pred (eds), *Violent Geographies: Fear, Terror and Political Violence* (London: Routledge, 200), pp. 111–132; R. Pain, R. 2009: 'Globalized fear? Towards an Emotional opolitics'*Progress in Human Geography* 33, 4, 466–486.

8. V.L. Henderson, 'Is There Hope for Anger? The Politics of Spatializing and (Re)Producing an Emotion', *Emotion, Space and Society* 1 (2008), 28–37.

9. H. Flam, 'Micromobilization and Emotions', in D. della Porta and M. Diani (eds), *The Oxford Handbook of Social Movements* (Oxford: Oxford University Press, 2015), pp. 264–276.

10. G.T. Eskew, 'Civil Rights Movement (United States)', in R.S. Powers, W.B. Vogele, C. Kruegler and R.M. McCarthy (eds), *Protest, Power, and Change* (New York: Garland Publishing, 1997), pp. 86–89.

11. G. Debord, *Society of the Spectacle* (Detroit: Black and Red, 1983).

12. See P. Routledge, 'A Spatiality of Resistances: Theory and Practice in Nepal's Revolution of 1990', in S. Pile and M. Keith (eds), *Geographies of Resistance* (London: Routledge, 1997), pp. 68–86.

13. Material for this section is drawn in part from P. Routledge, 'Sensuous Solidarities: Emotion, Politics and Performance in the Clandestine Insurgent Rebel Clown Army', *Antipode* 44/2 (2012), 428–452.

14. K. Klepto, and M. Up Evil, 'The Clandestine Insurgent Rebel Clown Army Goes to Scotland Via a Few Other Places', in D. Harvie, K. Milburn, B. Trott and D. Watts (eds), *Shut Them Down: The G8, Gleneagles 2005 and the Movement of Movements* (Leeds: Dissent, 2005), pp. 243–254.

15. S. Duncombe, *Dream: Reimagining Progressive Politics in an Age of Fantasy* (New York: New Press, 2007), p. 17.

16. K. Klepto, 'Making War with Love: The Clandestine Insurgent Rebel Clown Army', *City* 8/3 (2004), 403–411, see p. 407.

17. Quoted from: http://www.tacticalmediafiles.net/articles/3546/Clandestine-Insurgent-Rebel-Clown-Army-Manifesto;jsessionid=FCDD12E6EDoC3 9345D7ED7FE1B95F2A7 (accessed 12 September 2016); see also Klepto, 'Making War with Love', p. 407.

18. Klepto, 'Making War with Love'.

19. See M. Foucault, 'Governmentality', *Ideology and Consciousness* 6 (1979), 5–21.

20. Juris, 'Performing Politics'.

21. Duncombe, *Dream*, p. 131.

22. Juris, 'Performing Politics'.

23. B. Doherty, A. Plows and D. Wall, 'Environmental Direct Action in Manchester, Oxford and North Wales: A Protest Event Analysis', *Environmental Politics* 16 (2007), 805–825.

24. Klepto, 'Making War with Love', p. 406.

25. Duncombe, *Dream*, p.138.
26. Ahmed, *Cultural Politics of Emotions*, p. 15.
27. Quoted in Routledge, 'Sensuous Solidarities', p. 443.
28. H. Gorringe and M. Rosie, 'The Polis of "Global" Protest: Policing Protest at the G8 in Scotland', *Current Sociology* 56 (2008), 691–710.
29. D. della Porta and O. Fillieule, 'Policing Social Protest', in D. Snow, S. Soule and H. Kriesi (eds), *The Blackwell Companion to Social Movements* (Oxford: Blackwell, 2004), pp. 217–241.
30. J. Verson, 'Why We Need Cultural Activism', in Trapese Collective (ed.), *Do It Yourself: A Handbook for Changing the World* (London: Pluto Press, 2007), pp. 171–186.
31. G. Brown, 'Mutinous Eruptions: Autonomous Spaces of Radical Queer Activism', *Environment and Planning A* 39 (2007), 2685–2698.
32. G. Jobin-Leeds and AgitArte, *When We Fight We Win: Twenty-First-Century Social Movements and the Activists that are Transforming Our World* (New York: New Press, 2016).
33. D. Gould, *Moving Politics: Emotion and ACT UP's Fight against AIDS* (Chicago: University of Chicago Press, 2009).
34. J. Perini, 'Art as Intervention: A Guide to Today's Radical Art Practices', in Team Colors Collective, *Uses of a Whirlwind: Movement, Movements, and Contemporary Radical Currents in the United States* (New York: AK Press, 2010).
35. J. Wickman, 'Queer Activism: What Might That Be?' *Trickster* 4 (2010) (available at: http://trickster.net/4/wickman/1.html, accessed 31 January 2017).
36. M. Bronski, *A Queer History of the United States* (Boston: Beacon Press, 2011).
37. E. Addley and F. Perraudin, 'Sainsbury's Kiss-In Held after Lesbian Couple Told They Were "Disgusting"', *Guardian*, 15 October 2014 (available at: https://www.theguardian.com/world/2014/oct/15/sainsburys-brighton-kiss-in-protesters-lesbian-couple, accessed 31 January 2017).
38. S. Schulman, 'What Became of Freedom Summer?' *Gay and Lesbian Review* 11/1 (available at: http://www.glreview.org/article/what-became-of-freedom-summer/, accessed 27 September 2016).
39. Wickman, 'Queer Activism'.
40. Brown, 'Mutinous Eruptions'.
41. Jobin-Leeds and AgitArte, *When We Fight We Win*.
42. See 'Pussy Riot' (available at: https://en.wikipedia.org/wiki/Pussy_Riot, accessed 31 January 2017).
43. They were subsequently released by the Russian government, following international condemnation of their imprisonment and international solidarity actions.
44. Quoted from: http://pussy-riot.livejournal.com (accessed 9 September 2016).
45. Rist, quoted in M.-P. Kwan, 'Affecting Geospatial Technologies: Towards a Feminist Politics of Emotion', *Professional Geographer* 59/1 (2007), 22–34, quote from p. 30.

CHAPTER 8

1. I. Ortiz, S. Burke, M. Berrada and H. Cortés, 'World Protests 2006–2013', working paper, Initiative for Policy Dialogue and Friedrich-Ebert-Stiftung, 2013 (available at: http://www.fes-globalization.org/new_york/new-publication-world-protests-2006-2013/, accessed 31 January 2017).
2. Ibid.
3. EZLN (Ejercito Zapatista Liberacion National), *Critical Thought in the Face of the Capitalist Hydra, 1: Contributions of the Sixth Commission of the EZLN* (Durham, NC: Paper Boat Press, 2016).
4. Ibid.
5. Ortiz et al., 'World Protests 2006–2013'.
6. EZLN, *Critical Thought*.
7. See J. Rancière, *Dissensus: On Politics and Aesthetics* (London: Continuum, 2010).
8. M. Combes, 'Having the Last Word: Towards Paris 2015 – Challenges and Perspectives', *Ejolt*, 2015 (available at: http://www.ejolt.org/wordpress/wp-content/uploads/2015/09/EJOLT-6.42-52.pdf, accessed 31 January 2017).
9. N. Klein, *This Changes Everything: Capitalism vs. the Climate* (London: Penguin Books, 2014), p. 293.
10. The Invisible Committee, *For Our Friends* (South Pasadena, CA: Semiotext(e), 2015).
11. See D.A. Medina, 'Tribe Asks DOJ to Intervene in Escalating Dakota Access Pipeline Protests', *NBC News*, 24 October 2016 (available at: http://www.nbcnews.com/news/us-news/tribe-asks-doj-intervene-escalating-dakota-access-pipeline-protests-n671541, accessed 31 January 2017).
12. Cited in E. Wise, F. Barat and V. Cuervo, 'Indigenous Rights and the Fight for Life at Standing Rock', *ROAR*, 9 December 2016 (available at: https://roarmag.org/essays/standing-rock-interview-eryn-wise/, accessed 31 January 2017).
13. Ibid.
14. P. Marcuse, 'Keeping Space in Its Place in the Occupy Movements', *Progressive Planning* 191 (Spring 2012), 15–16.
15. J. Butler, *Notes Toward a Performative Theory of Assembly* (Cambridge, MA: Harvard University Press, 2015).
16. A. Feigenbaum, F. Frenzel and P. McCurdy, *Protest Camps* (London: Zed Books, 2013).
17. Butler, *Performative Theory of Assembly*.
18. Mauvaise Troupe Collective, 'Defending the Zad', Constellations, n.d. (available at: https://constellations.boum.org/spip.php?article143, accessed 31 January 2017).
19. B. Franks, 'Direct Action Ethic', *Anarchist Studies* 11/1 (2003), 13–41.
20. N. Bloch, 'The 2015 Creative Activist Awards', *Waging Nonviolence*, 1 January 2016 (available at: http://wagingnonviolence.org/feature/the-2015-creative-activist-awards/, accessed 31 January 2017).

21. S. Brennan, 'Notes from the Resistance: A Column on Language and Power', *Literary Hub*, 9 December 2016 (available at: http://lithub.com/notes-from-the-resistance-a-column-on-language-and-power/, accessed 31 January 2017).

22. K.D. Derickson and D. MacKinnon, 'Toward an Interim Politics of Resourcefulness for the Anthropocene', *Annals of the Association of American Geographers* 105/2 (2015), 304–312.

23. Y. Marom, 'What Really Caused the Implosion of the Occupy Movement: An Insider's View', *Alternet*, 23 December 2015 (available at: http://www.alternet.org/occupy-wall-street/what-really-caused-implosion-occupy-movement-insiders-view, accessed 17 March 2016).

24. Z. Grossman, 'Unlikely Alliances: Treaty Conflicts and Environmental Cooperation between Native American and Rural White Communities', *American Indian Culture and Research Journal* 29/4 (2005), 21–43.

25. See: http://www.americansagainstfracking.org.

26. See: http://www.ienearth.org.

27. R. Nunes, *Organization of the Organizationless: Collective Action after Networks* (Milton Keynes: Mute/Post-Media Lab, 2014), p. 21.

28. The Invisible Committee, *The Coming Insurrection* (Los Angeles: Semiotext(e), 2009).

29. M. Evans, *Artwash: Big Oil and the Arts* (London: Pluto Press, 2015).

30. N. Komami, 'BP to End Tate Sponsorship after 26 Years', *Guardian*, 11 March 2016 (available at: https://www.theguardian.com/artanddesign/2016/mar/11/bp-to-end-tate-sponsorship-climate-protests, accessed 31 January 2017).

31. S. Galeano, 'The Crack in the Wall: First Note on Zapatista Method', in EZLN, *Critical Thought*, p. 177.

32. For example, see: http://www.generativesomatics.org.

33. E. Sader, *The New Mole: Paths of the Latin American Left* (London: Verso Books, 2011).

34. J. Holloway, *Change the World Without Taking Power* (London: Pluto Press, 2010).

35. D. Mitchell, *The Right to the City* (London: Guildford Press, 2003).

36. A. Opel and G. Elmer, *Preempting Dissent* (Winnipeg: ARP Books, 2008).

37. E.O. Wright, *Envisioning Real Utopias* (London: Verso Books, 2010), pp. 303–307.

38. A. Cumbers, *Reclaiming Public Ownership: Making Space for Economic Democracy* (London: Zed Books, 2012).

39. S. Kishimoto, E. Lobina and O. Petitjean (eds), *Our Public Water Future* (Amsterdam: Transnational Institute, 2015).

40. J.A. Schertow, '15 Indigenous Rights Victories That You Didn't Hear About in 2015', *Intercontinental Cry*, 21 December 2015 (available at: https://intercontinentalcry.org/15-indigenous-rights-victories-didnt-hear-2015/, accessed 31 January 2017).

41. C.M. Schiavoni, 'The Contested Terrain of Food Sovereignty Construction: Toward a Historical, Relational and Interactive Approach', *Journal of Peasant Studies* 44/1 (2017), 1–32.

42. S. Beauregard, 'Food Policy for People: Incorporating Food Sovereignty Principles into State Governance', senior comprehensive report (Los Angeles: Urban and Environmental Policy Institute, Occidental College, 2009).

43. M. Menser, 'The Territory of Self-Determination: Social Reproduction, Agroecology, and the Role of the State', in P. Andree, M. Ayres, M.J. Bosia and M.-J. Massicotte (eds), *Globalization and Food Sovereignty: Global and Local Change in the New Politics of Food* (Toronto: University of Toronto Press, 2014), pp. 53–83.

44. H. Wittman, 'From Protest to Policy: The Challenges of Institutionalising Food Sovereignty', *Canadian Food Studies* 2/2 (2015), 174–182.

45. J. Fox, *Accountability Politics: Power and Voice in Rural Mexico* (New York: Oxford University Press, 2007).

46. J. Clover, *Riot. Strike. Riot: The New Era of Uprisings* (London: Verso, 2016).

47. A. Chakrabortty, 'For Real Politics, Don't Look to Parliament but to an Empty London Housing Estate', *Guardian*, 23 September 2014 (available at: https://www.theguardian.com/commentisfree/2014/sep/23/real-politics-empty-london-housing-estate, accessed 31 January 2017); see also: https://focuse15.org.

48. A. Cumbers, 'Constructing a Global Commons in, against and beyond the State' *Space and Polity* 19/1 (2015), 62–75.

49. R. Zibechi, *Dispersing Power: Social Movements as Anti-State Forces* (Edinburgh: AK Press, 2010).

50. M. Carter (ed.), *Challenging Social Inequality: The Landless Rural Workers Movement and Agrarian Reform in Brazil* (Durham, NC: Duke University Press, 2015).

51. See: http://righttothecity.org.

52. M. Sitrin and D. Azzellini, *They Cannot Represent Us: Reinventing Democracy from Greece to Occupy* (London: Verso Books, 2014).

53. S. Federici, *Caliban and the Witch* (New York: Autonomedia, 2004), p. 8.

54. S. Federici, 'Feminism and the Politics of the Commons in an Era of Primitive Accumulation', in Team Colors Collective (eds), *Uses of A Whirlwind: Movement, Movements, and Contemporary Radical Currents in the United States* (Edinburgh: AK Press, 2010), pp. 283–293, quote from p. 287.

55. Ibid., p. 288.

56. A. Negri and J. Roos, 'Toni Negri: From the Refusal of Labor to the Seizure of Power', *ROAR*, 18 January 2015 (available at: https://roarmag.org/essays/negri-interview-multitude-metropolis/, accessed 31 January 2017).

57. Nunes, *Organization of the Organizationless*, p. 33.

58. SEE: https://breakfree2016.org.

59. A. Gramsci, *Selections from the Prison Notebooks* (New York: International Publishers, 1971), pp. 330–334.

60. Sun Tzu, *The Art of War* (Boulder, CO: Shambhala Publications, 1988), p. 89.

61. J.A. Todd, 'Occupations, Assemblies and Direct Action – A Critique of "Body Politics"', *Red Pepper*, 22 August 2016 (available at: http://www.redpepper.org.uk/occupations-assemblies-and-direct-action-a-critique-of-body-politics/, accessed 31 January 2017).

62. S. Graham, 'Automated Repair and Backup Systems', in N.J. Thrift, A. Tickell, S. Woolgar and W.H. Rupp (eds), *Globalization in Practice* (Oxford: Oxford University Press, 2014), pp. 75–78.

63. '10 Facts Everyone Should Know about Anonymous', *Collective Evolution*, 22 October 2016 (available at http://www.collective-evolution.com/2016/10/22/10-facts-everyone-should-know-about-anonymous/, accessed 28 October 2016).

64. M. Sitrin, 'Social Reproduction: Between the Wage and the Commons', *Truthout*, 27 August 2016 (available at: http://www.truth-out.org/opinion/item/37358-social-reproduction-between-the-wage-and-the-commons, accessed 31 January 2017).

65. C. Mouffe, *Agonistics: Thinking the World Politically* (London: Verso Books, 2013), pp. 126–127, 133.

66. C. Delclós, 2015: 'The Logic of Overflow: From the Indignados to Podemos', *ROAR*, 18 December 2015 (available at: https://roarmag.org/essays/podemos-spain-fransisco-jurado-interviw/, accessed 31 January 2017).

67. N. Klein, *The Shock Doctrine* (London: Penguin Books, 2007).

68. Clover, *Riot*, p. 142.

69. S. Cossar-Gilbert, '#NuitDebout: A Movement Is Growing in France's Squares' *ROAR*, 6 April 2016 (available at: https://roarmag.org/essays/nuit-debout-republique-occupation/, accessed 31 January 2017).

70. See: www.beautifultrouble.org and www.activisthandbook.wordpress.com.

Index